V. Větvička
and P. Šíma

Evolutionary Mechanisms
of Defense Reactions

Springer Basel AG

Authors

Václav Větvička, Ph. D.
Assistant Professor
Dept. of Pathology
School of Medicine
Div. of Experimental Immunology
and Immunopathology
University of Louisville
Louisville, KY 20292
USA

Petr Šíma, Ph. D.
Senior Research Scientist
Dept. of Immunology
Institute of Microbiology
Academy of Science of the Czech Republic
Prague
Czech Republic

Die Deutsche Bibliothek – CIP-Einheitsaufnahme

Větvička, Václav:
Evolutionary mechanism of defense reactions / V. Větvička and
P. Šíma. – Basel ; Boston ; Berlin : Birkhäuser, 1998
    ISBN 978-3-0348-8835-6

Library of Congress Cataloging-in-Publication Data

Větvička, Václav.
    Evolutionary mechanisms of defense reactions / V. Větvička and
P. Šíma
            p.   cm.
    Includes bibliographical references and index.
    ISBN 978-3-0348-9793-8   ISBN 978-3-0348-8835-6 (eBook)
    DOI 10.1007/978-3-0348-8835-6
    1. Immune system--Evolution.   I. Šíma, Petr, 1942–
II. Title.
QR182.2.E94V48   1998
571.9′61--dc21   98-23992
                                                  CIP

©1998 Springer Basel AG
Originally published by Birkhäuser Verlag AG, in 1998
Softcover reprint of the hardcover 1st edition 1998

Printed on acid-free paper produced from chlorine-free pulp. TCF ∞
Cover design: Markus Etterich, Basel

ISBN 978-3-0348-8835-6

9 8 7 6 5 4 3 2 1

To Jana and Eva

A first step in analysis of any biological problem is descriptive and comparative study.

J. Huxley, 1958

Acknowledgement

The authors wish to thank Dr. Brian P. Thornton for his patient and tireless effort in helping to edit our book into proper English. We also express our deep gratitude to the authors and editors of excellent books published recently.

Finally, we thank and acknowledge Jana Vìtvièková and Eva Truxová, for their constant support, sympathy and understanding.

Louisville – Prague, January 1998

Contents

Introduction

At present, we do not fully understand at what stage of the evolution of living matter the first traces of defense reactions occurred. We even do not fully understand how and why immune systems reached their contemporary state in advanced vertebrates and man. It may be expected that in the near future these questions will be answered by comparative and developmental biology. Together with an extraordinary explosion of our knowledge about immunity of mammals including man, an increase in the interests concerning origin and development of immune mechanisms at lower stages of the phylogeny can be observed. The search for simple types of immune mechanisms in less complex but still evolutionary successful animals is promising and may contribute to better understanding of highly complex immune adaptive responses in mammals.

It is important to note that comparative and evolutionary immunology differs greatly from other branches of biomedical science. Apart from immunology and molecular biology, a specialist in this discipline has to be familiar with every detail of taxonomy, comparative anatomy, physiology, embryology, and even with the phyletic relationships of animals.

Probably no monography could deal with the entire animal kingdom, because, in many cases, the insights into questions about immune mechanisms of many animal groupings or phyla, and their possible evolutionary implications, are unknown or just now beginning to take shape. For the moment, our knowledge on such matters relies upon reconstructions of ideas that we have deduced from studies on members of relative taxa.

In recent years the theoretical trends of thought in evolutionary sciences diversified increasingly from Darwin's classical gradualism. Darwin truly believed in *"Natura non facit saltum,"* but in the light of new paleontological evidence this seems to be no longer valid. The evolution of living matter is neither proportional nor continual. In relatively quiet periods the majority of animal species stabilized, i.e., no substantial changes in morphology occurred. However, from the morphological structures we can deduce their functions and thus we can also assume that no substantial functional changes occurred either. It is more probable that in some species accumulation of new genetic information occurred through sexual recombination and/or by coupling of positive and negative mutations. This preadaptive genetic potential has had only a limited possibility of expression during unchanging environmental conditions. When dramatic changes of environment started taking place, the rapid extinction of non-preadapted species had to occur. Their place in the ecosystem was thus liberated for newly emerging species. The transitional, less stabilized forms of Metazoans were lost during this time of rapid adaptive radiation. The number of forms preserved by means of fossilization through their relatively short geological life was probably not very high. This hypothesis could explain not only missing links, but also a rapid and massive occurrence of new species which were endowed with a higher level of morpho-functional organization (Gould and Eldridge, 1977). Analogically, if we follow the evolution of defense mechanisms within one monophyletic group of animals, e.g., vertebrates, we shall discover that the immune profiles are stabilized in each vertebrate taxon. The transitional forms of immune mechanisms were excluded.

There is no function without structure, and there is no structure without a history. The immune strategy of every natural assemblage of animals represents an appropriate morphofunctional pattern determined by a common basic body plan which has emerged during the evolutionary history under the natural forces of environment in which these animals have radiated (Figs 1, 2). Only two fundamental patterns of immunity (Tab. 1) evolved: the constitutive (non-anticipatory) and the adaptive (anticipatory) immunity (Klein, 1989). To be endowed by the latter type means to be armed by three fundamental prerequisites:

1. The genetic equipment for generation of the rearranging receptors of which molecular configurations distinguish (and bind) between self and non-self, and non-self of one type with the other types.

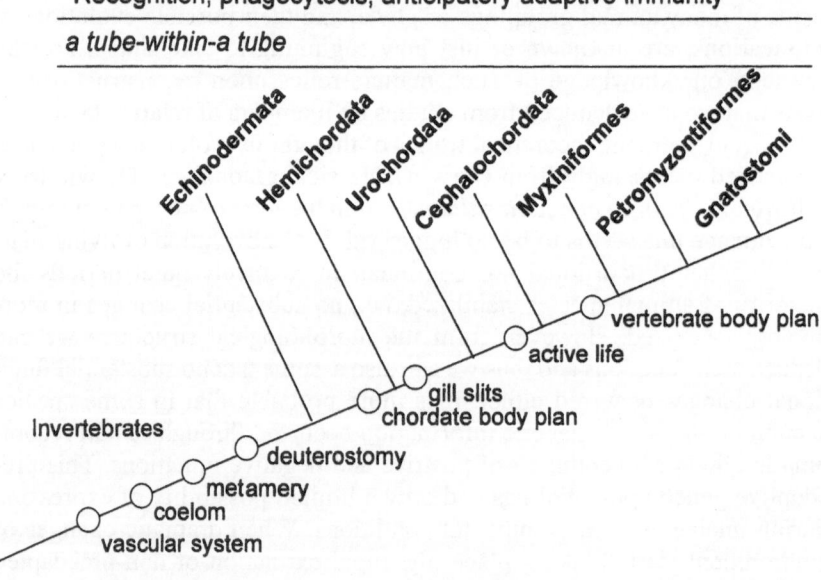

Figure 1. The morphoevolution in protostomian invertebrates.

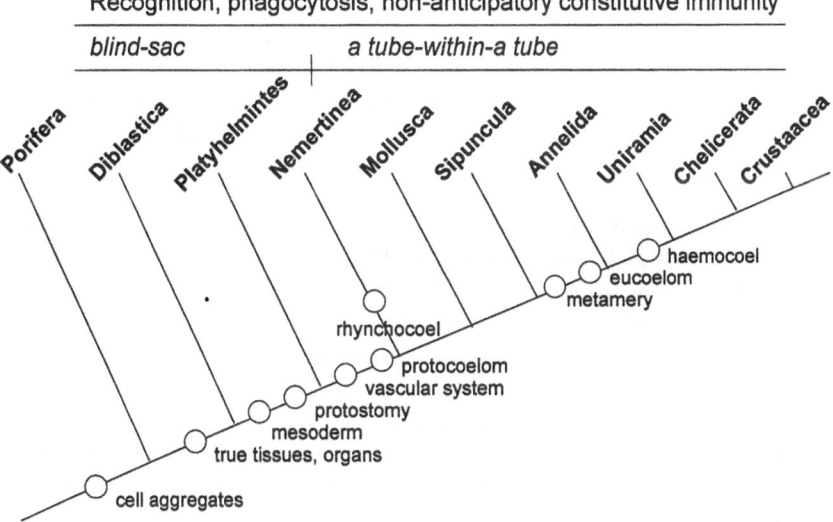

Figure 2. The morphoevolution in deuterostomes.

2. The cellular specialization ensuring the production of mutually collaborating cell clones manufacturing regulatory and specific defense substances.
3. The well-developed complex structures in which the microenvironment of these processes could be efficiently realized.

Whereas the vertebrates are capable of both types of immunity, the invertebrates seem to be devoid of the adaptive one, apart from rare instances

Table 1. The differences between two basic types of immune strategies, the constitutive and adaptive immunity

Constructive immunity	Adaptive immunity
Innate	Acquired
Non-anticipatory	Anticipatory
Nonspecific	Specific
Without memory	With memory
Recognition molecules	
Lectins	Ig family
Effectory cells	
Phagocytes	Lymphocytes
Effectory molecules	
Direct	Antibody-dependent
Cytotoxic	Classical
Non-Ig factors	Ig antibodies
Toxins	
Antibiotics	

in which particulate reactions exert some adaptive features, which nevertheless, are never underlined by the presence of rearranging molecular
mechanisms. Recently, some evidence from immunocomparative experiments suggests more kinship between the invertebrate and the vertebrate
immunities especially in the sense that in invertebrates there occur molecules and/or effectory mechanisms which seem to be precursors of those in
vertebrates (Cooper et al., 1992; Beck and Habicht, 1996).

Most evolutionists agree that all life, in all its variety, originated from
only one ancestral lifeform on Earth. This is based on the evidence that
there were no fundamental biochemical differences in the composition of
living matter whose existence relies upon the common principal functions
such as recognition, acceptance, storage and/or utilization, processing and
transmitting of biological information. From this point of view, hormones
may be defined as chemical messengers participating in the transmission
of this information within the organisms (Starling, 1905). In Kobayashi's
opinion, it was precisely the process of evolution and consolidation of the
system hierarchical hormonal regulations that assured more efficient stimulation of traditional metabolic pathways of manufacturing energy more
effectively (Kobaysahi, 1980). It is well known that 93% of the total of
enzyme activities are common to both the eukaryotes and prokaryotes
(Britten and Davidson, 1971). The basic metabolic pathways (glycolysis,
Krebs cycle, etc.) are essentially common to all living creatures. The evolution of metazoans clearly shows that the ancestral forms with a characte-
ristically sessile way of life are gradually substituted by creatures actively
seeking food. This process required, however, the origination of structures
and organs that would ensure more developed locomotory functions
(Clark, 1964; Romer, 1971). This increasing locomotory activity has been
shown to be a predominant factor in the development of the more evolutionary advanced metazoan supporting locomotory structures such as the
coelom and segmentation of the body, which induced the development of
appendages in the protostomia phyla, as well as metamery, notochord, and
appendages in the chordates, and tube feet in the echinoderms.

Thus, these evolutionary changes in body structures allowed these creatures to abandon their passive style of life; on the other hand, this form of
evolution required a considerable increase in their energy supply. More
energy sufficient food was required for their more active life, but it could
not be fully exploited without a fundamental improvement of their metabolism. For this reason, the need for changes was not in their metabolic
structures, but rather in their metabolic processes. This meant that the ancestral metabolic system was no longer sufficient, but principially it remained unchanged – instead it started to be coordinated by new regulatory
molecules: hormones. In this sense, hormones can be considered to be products of biological progress.

New, quantitatively higher levels in the evolution of living matter
emerged in key periods of the Earth's history, in periods of so-called "evo-

lutional revolutions." The first of these periods occurred in early Precambrium when the first eukaryotes emerged, and had to maintain their own integrity against prokaryotes (Schopf, 1972). Eukaryotic cells and primitive eukaryotic organisms possessed their own recognition mechanisms capable of self/non-self recognition (Holtfreter, 1939). Present-day protozoans can serve as an example (Tartar, 1970). Since that time, we are able to use the term immunity, which together with metabolism, reproduction and other attributes are among the major attributes of all eukaryotic organisms (Burnet, 1970).

The next period, "Cambrian Explosion" occurred approximately 600 million years ago. In that period all basic forms of metazoans emerged. The multicellularity entailed strict requirements for the coordination of specialized cell populations. The requirements were greater when these populations began to become differentiated into tissues and organs. The evolutional needs of this integration led to the origin of basic homeostatic nervous, endocrine and defensive systems. The emergence of defenses and of adaptive immunity were unquestionably the result of such an increasing organizational hierarchy. Not only the interconnection and integration of distant parts of multicellular organisms, but also other eventualities could support the emergence of homeostatic systems, especially that of defense.

When searching for the evolution of immunocompetent structures, we must be aware that the structure is a *condition sine qua non* of the presence of function. This means that the immune mechanisms belonging to the members of each animal phylum are determined by the fundamental body plan characteristic for the particular assemblage. In the evolutionary context, the basic body pattern represents the most convenient evolutionary compromise between the morphofunctional possibilities determined by the genome and external selective pressures. It must be kept in mind that the membership of the animals in some natural assemblage (like a phylum) is conditioned by common ancestry and basic body pattern from which only non-substantial modifications have developed. Simultaneously, these eventualities determine the type of the immune strategy, which is also common for all taxonomically related animals. The emergence of a new, more advanced body organization accompanied by evolutionary innovations has always been preceded by the emergence of a new basic body plan. Analogically, all innovations of immune mechanisms had to be associated with morphofunctional development of structural novelties which have accumulated during the phylogeny (Valentine et al., 1996). In the following overview we have briefly summarized the major immune characteristics of the main animal taxa from poriferans to agnathans (Tab. 2). The gnathostomean classes, the chondrichthyans and osteichthyans are not included, for their members are endowed by all fundamental attributes of true adaptive immunity known in endotherms.

Table 2. The survey of fundamental body plans of metazoan animals in respect to main immune
phenomena

Animals at tissue grade of construction (Parazoa)

Irregular or radial, cell-aggregate basic body plan

The poriferans
Highly discriminative and collaborative type of cellular immunity with main cell type
(archaeocyte) and a row of accessory types (pinacocyte, choanocyte, spherulocyte, col-
lencyte, spherulocyte); immune reactions such as phagocytosis, cytotoxicity, nonfusion, transplantation reaction in which the humoral substances, the lectins, toxic
lysins, and toxins participate; some animals are endowed with the inducible, low specific immunological memory.

Animals at diploblastic grade of construction (Eumetazoa)

Radially symmetrical sac-like basic body plan

The coelenterates and ctenophorans
Phagocytosis and cytotoxicity are main reactions of cellular immunity in which amoebocytes and granulocytes participate as main cell types; the lectins, toxic lysins, and
toxins often secreted in mucus are vectors of humoral immunity; in some species a low
specific inducible immunological memory exists.
Characteristic phenomenal regeneration.

Animals at the tribloblastic grade of construction

Acoelomates

Protostomes

Bilaterally symmetrical sac-like basic body plan

The platyhelminths
Lack of substantial information (neoblast is supposed to be a vector of cell immunity;
humoral defense substances are lectins, toxic lysins, and toxins).
Characteristic rapid effective regeneration.

A tube-within-a-tube basic body plan

The nemerteans
Main types of cell immunity are the macrophage-like and lymphocyte-like cells (phagocytosis); lectins, toxic lysins, and toxins are factors of humoral immunity; rejection
of xenografts with short memory.
First example in the phylogeny of a true closed circulatory system and a coelom-like
cavity; the rhynchocoel appears.

Pseudocoelomates
The aschelminths
Cellular and humoral vectors of immunity unknown; without circulatory system.

Schizocoelous eucoelomates

The sipunculans
Cell immunity is represented by phagocytosis and cytotoxic reactions in which the granulocytes collaborate with ciliated urns; only recognition of self and non-self; lysins
and agglutinins in humoral response. Without circulatory system.

Schizocoelous metamerized eucoelomates

The echiurans
Lack of substantial information; only antibacterial humoral response. Segmentation
during embryogenesis, closed vascular system.

Table 2. (continued)

Schizocoelous metamerized eucoelomates

The annelids

Many types of coelomocytes (amoebocytes) collaborate with specialized types of accessory eleocytes in phagocytosis, cytotoxicity, and xenograft rejection; in humoral immunity (molecules with antibacterial, antiviral, lysins, agglutinins, opsonins), cytotoxic, and primitive cytokine-like activities; well-developed, relatively high specific immunological memory.

True closed circulatory system. First invertebrate animals having a specific anamnestic immune response, a highly sophisticated system of mutually collaborating cellular and humoral defense components.

The arthropods

Plenty of main immune cell types (hemocyte, granulocyte, plasmatocyte) collaborating with accessory hyaline cells, fixed phagocytes, and cystocytes in phagocytosis, encapsulation, cytotoxicity, and in some cases in allo- and xenograft rejection; the cercopins, attacins, diptericidins, defensins, lysins, lectins, prophenoloxydase system are humoral immune substances; well-developed immunological memory only in some taxa.

Open circulatory system, mixocoel. The highly discriminative, more or less specific, sometimes anamnestic reponse, well-developed cooperation.

The lophophorates

No information available.

The molluscs

The hemocytes, hyalinocytes, granulocytes, and leukocytes cooperating with serous cells and fixed cells in phagocytosis, cytotoxicity, and rejection of xenografts; humoral vectors are lysins, lectins, and opsonins; without immunological memory.

Primary body cavity transformed into hemocoel, coelom reduced.

Deuterostomes

Enterocoelous eucoelomates

The echinoderms

In phagocytosis, encapsulation, clotting, and graft rejection the phagocytic amoebocyte collaborates with accessory types, represented by the modifications of spherule cell, vibratile cell, and hemocyte; accelerated rejection of second-set allo- and xenografts; analogy to vertebrate NK cells; in humoral response, the presence of antibody-like factor, IL-1-like factor lysins, lectins, opsonins, and receptors for complement components have been proved; highly discriminative specific transplantation memory.

Coelom specialized into hemocoelic and vascular system. First animals with many fundamental morphofunctional features of vertebrate adaptive immunity.

The cephalochordates

Lack of information, only phagocytosis realized by coelomocytes detected.

The urochordates

Cooperating leucocytes, lymphocytes, and hemoblasts as main cell types with vacuolated cells in phagocytosis, encapsulation, allogenic rejection, colony specificity, programmed senescence, inflammation, and recognition regeneration; transplantation reaction is governed by a MHC-like locus mechanism similar to that in vertebrates; lectins, bactericidins, molecules of Ig superfamily (Thy 1, Lyt 1,2,3-like) opsonins, antiviral substances, and IL-like activities in humoral response; no information on immunological memory.

Coelom reduced to the pericardial cavity, hemocoelic vascular system well-developed. Hematogenic tissue in discrete structures in the branchial region and along the digestive tract regarded to be homologic with vertebrate hemopoletic tissues, the lymphocyte regarded as homologic to vertebrate lymphoid cell, structural basis for macrophage-phagocytic system of vertebrates, continual renewal of immunocompetent cells.

Table 2. (continued)

The agnathans
True lymphocytes, macrophages, and various kinds of leucocytes cooperate with macrophages in immune recognition phagocytosis, allogenic rejection, and inflammation processes; transplantation reaction is governed by a MHC-like locus; lectins, bactericidins, molecules of Ig superfamily (Thy 1, Lyt, IL-like molecules), components of complement cascade, opsonins, and antiviral substances are engaged in humoral response; presence of specific immunological memory is doubtful.
Hematogenic tissue in discrete structures in the branchial region and along the digestive tract regarded to be homologic with those of jaw vertebrates.

References

Beck, G. and Habicht, A.S. (1996) Immunity and the invertebrates. *Scientific American* 274:42–46.

Britten, R.J. and Davidson, E.H. (1971) Repetitive and non-repetitive DNA sequences and a speculation on the origins of evolutionary novelty. *Quart. Rev. Biol.* 46:111–133.

Burnet, F.M. (1970) *Immunological Surveillance*. Pergamon Press, Oxford.

Clark, R.B. (1964) *Dynamics in Metazoan Evolution. The Origin of the Coelom and Segments*. Claredon Press, Oxford.

Cooper, E.L., Rinkevich, B., Uhlenbruck, G. and Valembois, P. (1992) Invertebrate immunity: another viewpoint. *Scand. J. Immunol.* 35:247–266.

Gould, S.J. and Eldridge, N. (1977) Punctuated equilibria: the tempo and mode of coevolution reconsidered. *Pathobiology* 3:115–151.

Holtfreter, J. (1939) Gewebeaffinität, ein Mittel der embryonalen Formbildung. *Arch. Exp. Zellforsch. Gewebezucht.* 23:169–183.

Kobayashi, H. (1980) Evolution of the metabolic endocrine system: its phylogenetic significance. *In:* S. Ishii (ed.): *Hormones, Adaptation and Evolution*, Japan Sci. Soc. Press, Tokyo, pp 15–30.

Klein, J. (1989) Are invertebrates capable of anticipatory immune responses? *Scand. J. Immunol.* 29:499–505.

Romer, A.S. (1971) *The Vertebrate Body*. W.B. Saunders, Philadelphia.

Schopf, J.W. (1972) Precambrian paleobiology. *In:* C. Ponamperuna (ed.): *Exobiology*, North-Holland Publishing Co., Amsterdam, pp 16–34.

Starling, E.H. (1905) The chemical correlation of the functions of the body. *Lancet* II:339–342.

Tartar, V. (1970) Transplantation in *Protozoa. Transpl. Proc.* 2:183–190.

Valentine, J.V., Erwin, D. H. and Jablonski, D. (1996) Developmental evolution of metazoan bodyplans: the fossil evidence. *Develop. Biol.* 173:373–381.

Animals at the cell-aggregate body organization

Porifera

Porifera (sponges) are the most primitive metazoans. Approximatelly 5000 species of sponges are marine sessile animals, only some 150 live in fresh water. Four classes of *Porifera* are distinguished: the *Calcarea (Calcispongiae)*, the *Hexactinellida (Hyalospongiae)*, the *Demospongiae*, and the *Sclerospongiae* (Bergquist, 1978; Müller et al. 1994b).

The sponge body is built according to a radially symmetrical or irregular cell aggregate, and diploblastic basic body plan. The architecture of the sponge body is unique as it has a system of water canals, which is not seen within the metazoan phyla. The sponges were for a long time considered to be plants until the internal water channels were discovered. These animals do not possess a vascular system, or any distinct organs including a mouth and digestive system, and the nervous system is absent, too.

Sponge cells never form true tissues: separately imbedded in the extracellular matrix the cells generally display a high degree of independence, and morphologically and functionally are considerably specialized. Sponge cells are assembled in two layers forming an outer and an inner epithelium surrounding a central cavity named spongocöele. The layers are separated from each other by the mesophyl (mesoglea) into which cells of outer epithelium move, forming a mesenchyme-like tissue. Various types of amoeboid cells freely wander mainly in mesoglea, but also through other parts of a body, and are significant elements of the sponge cellular immunity. Traditionally, immune reactions of sponges have been described as phagocytosis, wound healing, agglutination, graft rejection and cytotoxicity. More recently, sponges' production of chemical defense molecules is being used as another example of their rather sufficient defense system.

Humoral immunity

Pfeifer et al. identified a TNF-like activity in the xenograft of the sponge *Geodia cydonium* (1992) (Fig. 1). Using an ELISA assay, the TNF-like activity was detected from day 2 after grafting with peak levels at days 4 and 5. By day 1, an inhibitory aggregation factor gp27, a product of the gp180 lectin receptor, was formed. The authors concluded that gp27 induced TNF-like production, resulting in destruction and dissolution of the xenograft (Fig. 2).

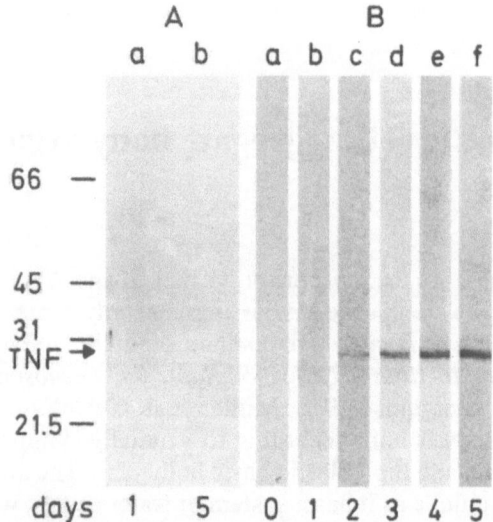

Figure 1. An extract from a 1- or 5-day-old xenograft from *Geodia cydonium* was electro-phoresed. Proteins were transferred to nitrocellulose sheets (A) and incubated first with anti-TNF antibodies followed by incubation with peroxidase-conjugated secondary antibodies. (B) Appearance of TNF-like activity in xenografts in dependence of a 0- to 5-day grafting periods. (From Pfeifer et al., 1992, with permission.)

The first demonstration of the agglutinating effects in sponges is almost 60 years old (Galtsoff, 1929). Since then, agglutinating and hemagglutinat-ing properties of cell extracts were found in almost all sponges used. (For more detailed review, see Ey and Jenkin, 1982.) Some studies suggested that lectins or lectin-like molecules with agglutinating properties are more involved in adhesion than in real defense. A carbohydrate-carbohydrate interaction of a novel acidic glycan found in *Microciona prolifera* mediates sponge cell adhesion (Misevic and Burger, 1993; Spillmann et al., 1993).

The cloning of two lectins of *Geodia cydonium* showed strong similari-ties with the carbohydrate-binding sites of vertebrate-type lectins (Gamu-lin et al., 1994; Nakao et al., 1993). Similarly, ubiquitination of proteins is a critical step in the controlled degradation of many polypeptides. The polyubiquitin cDNA was isolated and characterized from the sponge *Geo-dia cydonium*. A comparison of the amino acid composition revealed over 93% homology to those of higher organisms (Pfeifer et al., 1993). Ubi-quitin is involved in processes which control cell-matrix adhesion in spon-ges (Müller et al., 1994a). Subsequently, the same group isolated, cloned and characterized a receptor tyrosine kinase of class II, and deduced amino acid sequence with two characteristics: the tyrosine kinase domain and an immunoglobulin-like domain. The latter part displayed a high homology to the vertebrate type immunoglobulin domain. These results together strongly suggested that binding domains of such adhesion molecules are

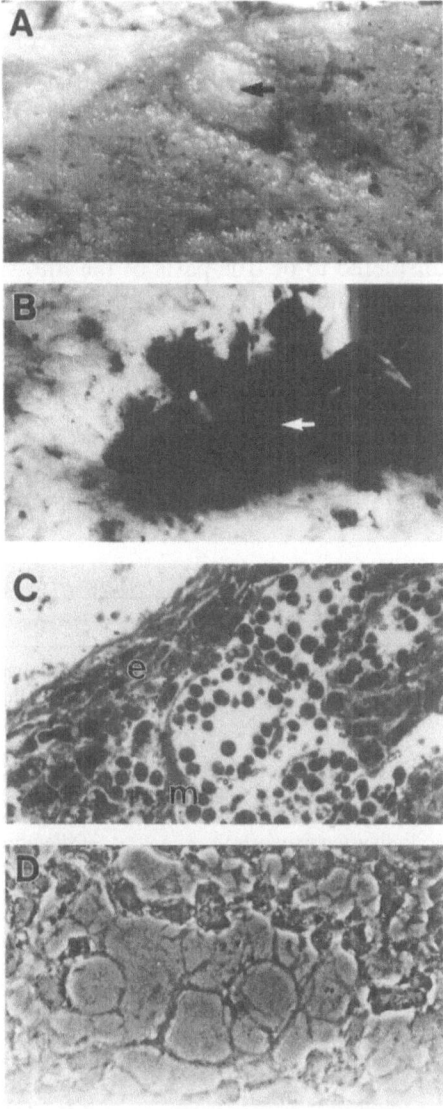

Figure 2. Introduction of cytotoxicity of *Geodia cydonium* tissue by human TNF-α injection. At 1 day (the necrotic tissue is labelled with an arrow) (A), or 2 days (B) (arrow: fullblown necrosis) the animal was cross-sected. Samples from control tissue (C) and of necrotic tissue (D). Magnification: A and B original size, C and D × 200. (From Pfeifer et al., 1992, with permission.)

not an achievement of evolutionary higher creatures, but exist already in sponges.

Marine invertebrates are currently the focus of an intense search for new cytotoxic, immunomodulating or antineoplastic agents. In sponges, several biologically active natural compounds have been found. Their list is summarized in Table 1 and reviewed in (Proksch, 1994). Readers seeking a detailed chemical structure of these compounds should check the original references listed in that table. It is not quite clear if these chemical molecules should be considered to be true parts of the immune system (understood as a defense of the integrity of an organism), or more or less defense against predators or competition for space. The second possibility is supported by findings that the highest incidence of these compounds is found in habitats such as coral reefs, where intense competition and feeding pressures exist (Proksch, 1994).

Table 1. Bioactive compounds of *Porifera*

Compound	Species	Reference
Latrunculin A, B	*Latrunculia magnifica*	(Kashman et al. 1980)
Heteronemin		(Proksch, 1994)
Scalardial		(Proksch, 1994)
Manoalide		(Proksch, 1994)
Seco-manoalide		(Proksch, 1994)
Siphonodictidine	*Siphonodictyon coralliphagum*	(Sullivan et al., 1981)
1-Methyladenine	*Aplysilla glacialis*	(Bobzin and Faulkner, 1992)
Aerothionine	*Aplysina fistularis*	(Thompson, 1985)
Aeroplysinin	*Verongia aerophoba*	(Teeyapant et al., 1993a)
Dibromotyrosine	*Verongia aerophoba*	(Teeyapant et al., 1993b)
Dienone	*Verongia aerophoba*	(Teeyapant et al., 1993b)
Bastadin 8 and 9	*Ianthella basta*	(Miao et al., 1990)
Sesterterpenes	*Luffariella*	(Konig et al., 1992)
Halichondin B	*Lissodendoryx*	(Flam, 1994)
Halistatin-1	*Phakellia carteri*	(Pettit et al., 1993)
Xestobergsterol A, B	*Xestospongia bergquistia*	(Takei et al., 1993)
Popolohuanone E	*Dysidea*	(Carney and Scheuer, 1993)
Purpurone	*Iotrochota*	(Chan et al., 1993)
Spongistatin-1	*Spongia*	(Bai et al., 1993)
Sesterterpenes	*Spongia*	(He et al., 1994)
Proteinases	*Eohydatia mulleri*	(Avenirova et al., 1992)
Sesquiterpenoids	*Reniera fulva*	(Casapullo et al., 1993)
Raspacionin B	*Raspaciona aculeata*	(Cimino et al., 1993)
Euryspongiols	*Euryspongia*	(Dopeso et al., 1994)
Desacetylaltohyrtin A	*Hyrtios altum*	(Kobayashi et al., 1993a,b)
Trichoharzin	*Micale cecilia*	(Kobayashi et al., 1993c)
9,11 Secosterol	*Gersenua fruticosa*	(Koljak et al., 1993)
Phloeodistines A1–A7	*Phloeodisctyon*	(Kouranvlefoll et al., 1994)
Hexacyclin bisguanidine	*Stylotella agminata*	(Kinnell et al., 1993)
Toxadocial A, B, C	*Toxadocia cylindrica*	(Nakao et al., 1993a,b)
Axinastatins 2,3	*Axinella*	(Pettit et al., 1994)

Cellular immunity

The intracellular digestion in these microphagous animals is conditioned by their strong power of phagocytosis. All sponge cells are able to engulf particles. Larger particles are phagocytosed by outer pinacocytes, whereas sizes lesser than 1 µm are phagocytosed by choanocytes forming inner epithelia, and by wandering amoebocytes, serving as nutritional and waste transports, as well as for the food storage (Weissenfels, 1976; Willenz, 1980). The primary digestive function, phagocytosis, is, in the instance of poriferans, maximally utilized in the first line of defense against invading non-self molecules (van de Vyver, 1981; Francis and Poirrier, 1986). There is an archeocyte, a type of free amoeboid cell, which is a conspicuous phagocytic cell named "a macrophage of sponges". The archeocytes are also considered to be polypotent stem cells from which other cell types arise (Borojevic, 1966). Also the specialized collencytes, of which the main role is production of collagen and formation of tissue wall during encapsulation, are capable of phagocytosis to a lesser extent. All cells filled by phagocyted material migrate throughout the mesophyl towards the water channels. Then, after crossing the pinacoderm, they are removed by passing into the water stream leaving the body.

Phagocytosis is conditioned by the recognition of self and non-self. The recognition phenomena in sponges were described already in 1907 (Wilson, 1907). Only the cells originating from the one species-type reaggregate; if the cells from genetically different animals are intermingled, chimeras never develop, and again, original sponge species reaggregate. Reaggregation reaction is composed of three different steps: the cellular contact stage followed by cell adhesion, and finally, the histiotypic rearrangement of cells composing new individuals (Johnston and Hildemann, 1982). Since 1963, when the soluble "aggregation factor" was described, a number of low- and high-molecular weight substances participating in recognition and cellular adhesion have been isolated, from which many exert the agglutination properties (Ey and Jenkin, 1982). These substances are suggested to posses ubiquitous lectin characters which play a more significant role in cellular adhesion, aggregation, and morphogenesis, than in defense (Vasta, 1991).

Allogeneic reaction of sponges has been studies in details by Hildemann and his group (Hildemann et al., 1979, 1980). Based on the presence of the elevated rate of second-set grafts, he considered this reaction to be at least partly comparable to the adaptive immunity of higher vertebrates (Evans et al., 1980). The cells responsible for the rejection seems to be the archaeocytes because of their cytotoxic reaction (Buscema and van de Vyver, 1984). Moreover, in the rejection response the collencytes which form the collagen barrier between allograft and self tissue also participate. Spherulocytes collaborate by secretion of hemagglutinating lectins (Bretting and Konigsmann, 1979). The cytotoxic power of the archeocytes during rejection resembles NK cell activity in vertebrates. Killing activities of both types of these cells are not dependent on the previous sensitization (Buscema and van de Vyver, 1984).

Mukai and Shimoda (1986) studied histocompatibility in four species of freshwater sponges *Ephydatia muelleri*. Gemmules obtained from two different localities were used, and it was demonstrated that all combinations of specimens coming from one locality were fusible, but those between samples from different localities were rejected. Contrary to these findings, in chimeric sponges cultured in the same location of their origin, the rejection reaction occurred even between individual sponges in the same location, but only after more than 10 days. Therefore, this reaction was probably missed in previous experiments (Mukai, 1992). Humphreys studied allorecognition in the marine sponge *Microciona prolifera* at a cellular level by apposing individual-specific aggregates from two different sponges (Humphreys, 1994). In about 4 to 6 h after contact, a yellow line formed as a boundary between the two aggregates. These data suggest a rapid allorecognition without prior sensitization. This system surely deserves further studies, because it might be an excellent model for examination of the development of self/non-self recognition on a very simple level. Conclusively, the sponge transplantation reaction exerts some features of adaptive allogeneic reaction of vertebrates like short-time immunological memory (Hildemann et al., 1980), more rapid second-set rejection (Evans et al., 1980), cytotoxicity, and capability to distinguish between inert and living material (Buscema and van de Vyver, 1985).

Conclusions

Basic immune mechanisms of sponges are phagocytosis and encapsulation of which the dominant cellular type is an archeocyte also endowed with a strong cytotoxic effect. Despite the primitive character of sponges, the cellular cooperation processes during immune reactions are typical features comparable in their efficacy with those in vertebrates. The sponges realize quasi-immune phenomena connected with a specific short-time type of immunological memory. Besides humoral immune factors of lectin type, they produce several bioactive substances that regulate their interrelationships with other micro- and macro-coinhabitants of their biocenose.

References

Avenirova, E.L., Rudenskaya, G.N., Filipova, I.Y. and Stepanova, V.M. (1992) Proteinases from gemmules of a freshwater sponge. *Biochemistry – Russia* 57:841–847.
Bai, R.L., Cichacz, Z.A., Herald, C.L., Pettit, C.L. and Hamel, E. (1993) Spongistatin-1, a highly cytotoxic, sponge-derived, marine natural product that inhibits mitosis, microtubule assembly, and the binding of vinblastine to tubulin. *Mol. Pharmacol.* 44:757–766.
Bergquist, P.R. (1978) *Sponges*. University of California Press, Berkeley.

Bobzin, S.C. and Faulkner, D.J. (1992) Chemistry and chemical ecology of the Bahamian sponge *Aplysilla glacialis*. *J. Chem. Ecol.* 18:309–332.

Borojevic, R. (1966) Étude expérimentale de la différenciation des cellules de l'éponge au cours de son développement. *Develop. Biol.* 14:130–153.

Bretting, H. and Konigsmann, K. (1979) Investigation on the lectin-producing cells in the sponge *Axinella polyploides* (Schmidt). *Cell Tissue Res.* 201:487–497.

Buscema, M. and van de Vyver, G. (1984) Cellular aspects of alloimmune reaction in sponges of the genus *Axinella*. I. *Axinella polyploides*. *J. Exp. Zool.* 229:7–17.

Buscema, M. and van de Vyver, G. (1985) Cytotoxic rejection of xenografts between marine sponges. *J. Exp. Zool.* 235:297–308.

Carney, J.R. and Scheuer, P.J. (1993) Popolohuanone-E. A topoisomerase – II inhibitor with selective lung tumor cytotoxicity from the Pohnpei sponge *Dysidea* sp. *Tetrahedron Lett.* 34:3727–3730.

Casapullo, A., Minale, L. and Zollo, F. (1993) Paniceins and related sesquiterpenoids from the mediterranean sponge *Reniera fulva*. *J. Nat. Products* 56:527–533.

Chan, G.W., Francis, T., Thureen, D.R., Offen, P.H., Pierce, N.J., Westley, J.W., Johnson, R.K., Faulkner, D.J. and Faulkner, D.R. (1993) Purpurone, an inhibitor of ATP citrate lyase – a novel alkaloid from the marine sponge *Iotrochota* sp. *J. Org. Chem.* 58:2544–2546.

Cimino, C., Crispino, A., Madaio, A., Trivellone, E. and Uriz, M. (1993) Raspacionin-B, a further triterpenoid from the mediterranean sponge *Raspaciona aculeata*. *J. Nat. Products* 56:534–538.

Dopeso, J., Quinoa, E., Riguera, R., Debitus, C. and Bergquist, P.R. (1994) Euryspongiols – 10. New highly hydroxylated 9,11-secosteroids with antihistaminic activity from the sponge *Euryspongia sp.* Stereochemistry and reduction. *Tetrahedron* 50:3813–3828.

Evans, C.W., Kerr, J. and Curtis, A.S.G. (1980) Graft rejection and immune memory in marine sponges. *In:* M.J. Manning (ed.): *Phylogeny of Immunological Memory*, Elsevier/North-Holland Biomed. Press, Amsterdam, pp 27–34.

Ey, P.L. and Jenkin, C.R. (1982) Molecular basis of self/non-self discrimination in the invertebrata. *In:* N. Cohen and M.M. Sigel (eds): *The Reticuloendothelial System, Vol. 3*, Plenum Press, New York, pp 321-391.

Flam, F. (1994) Chemical prospectors scour the seas for promising drugs. *Science* 266:1324–1325.

Francis, J.C. and Poirrier, M.A. (1986) Particle uptake in two fresh-water sponge species, *Ephydatia fluviatilis* and *Spongilla alba (Porifera: Spongillidae). Trans. Am. Microsc. Soc.* 105:11–20.

Galtsoff, P.S. (1929) Heteroagglutination of dissociated sponge cells. *Biol. Bull.* 57:250–260.

Gamulin, V., Rinkevich, B., Schacke, H., Kruse, M., Müller, I.M. and Müller, W.E.G. (1994) Cell adhesion receptors and nuclear receptors are highly conserved from the lowest metazoa (Marine sponges) to vertebrates. *Biol. Chem. Hoppe-Seyer* 375:583–588.

He, H.Y., Kulanthaivel, P. and Baker, B.J. (1994) New cytotoxic sesterterpenes from the marine sponge *Spongia sp. Tetrahedron Lett.* 35:7189–7192.

Hildemann, W.H., Johnston, I.S. and Jokiel, P.L. (1979) Immunocompetence in the lowest metazoan phylum: Transplantation immunity in sponges. *Science* 204:420–422.

Hildemann, W.H., Bigger, C.H., Jokiel, P.L. and Johnston, I.S. (1980) Characteristics of immune memory in invertebrates. *In:* M.J. Manning (ed.): *Phylogeny of Immunological Memory*, Elsevier/North-Holland Biomed. Press, Amsterdam, pp 9–14.

Humphreys, T. (1994) Rapid allogeneic recognition in the marine sponge *Microciona prolifera*. Implications for evolution of immune recognition. *Annals New York Acad. Sci.* 712:342–345.

Johnston, I. and Hildemann, W.H. (1982) Cellular defense systems of the Porifera. *In:* N. Cohen and M.M. Sigel (eds): *The reticuloendothelial System. Vol. 3.* Plenum Press, New York, pp 37–57.

Kashman, Y., Groweiss, A. and Shmueli, U. (1980) Latrunculin, a new 2-thiazolidinone macrolide from the marine sponge *Latrunculia magnifica*. *Tetrahedron Lett.* 21:3629–3632.

Kinnell, R.B., Gehrken, H.P. and Scheuer, P.J. (1993) Palauamine – a cytotoxic and immunosuppressive hexacyclic bisguanidine antibiotic from the sponge *Stylotella agminata*. *J. Am. Chem. Soc.* 115:3376–3377.

Kobayashi, M., Aoki, S., Sakai, H., Kawazoe, K., Kihara, N., Sasaki, T. and Kitagawa, I. (1993a) Altohyrtin-A, a potent anti-tumor macrolide from the Okinawan marine sponge *Hyrtios altum*. *Tetrahedron Lett.* 34:2795–2798.

Kobayashi, M., Aoki, S., Sakai, H., Kihara, N., Sasaki, T. and Kitagawa, I. (1993b) Altohyrtin-B and altohyrtin-C and 5-desacetylaltohyrtin-A. Potent cytotoxic macrolide congeners of altohyrtin-A, from the Okinawan marine sponge *Hyrtios altum. Chem. Pharmaceut. Bull.* 41: 989–991.

Kobayashi, M., Uehara, H., Matsunami, K., Aoki, S. and Kitagawa, I. (1993c) Trichoharzin, a new polyketide produced by the imperfect fungus *Trichoderma harzianum* separated from the marine sponge *Micale cecilia. Tetrahedron Lett.* 34:7925–7928.

Koljak, R., Pehk, T., Jarving, I., Liiv, M., Lopp, A., Varvas, K., Vahemets, A., Samel, N. and Lille, U. (1993) New antiproliferative 9,11-secosterol from solf coral *Gersemia fruticosa. Tetrahedron Lett.* 34:1985–1986.

Konig, G.M., Wright, A.D. and Sticher, O. (1992) Four new antibacterial sesterpenes from a marine sponge of the genus *Luffariella. J. Nat. Products* 55:174–178.

Kouranvlefoll, E., Laprevote, O., Sevenet, T., Montagnac, A., Pais, M. and Debitus, C. (1994) Phloeodictines A1-A7 and C1-C2. Antibiotic and cytotoxic guanidine alkaloids from the New Caledonian sponge, *Phloeodictyon sp. Tetrahedron* 50:3415–3426.

Miao, S., Andersen, R.J. and Allen, T.M. (1990) Cytotoxic metabolites from the sponge *Ianthella basta* collected in Papua New Guinea. *J. Nat. Products* 53:1441–1446.

Misevic, G.N. and Burger, M.M. (1993) Carbohydrate-carbohydrate interactions of a novel acidic glycan can mediate sponge cell adhesion. *J. Biol. Chem.* 268:4922–4929.

Mukai, H. (1992) Allogeneic recognition and sex differentiation in chimeras of the freshwater sponge *Ephydatia muelleri. J. Exp. Zool.* 264:298–311.

Mukai, H. and Shimoda, H. (1986) Studies on histocompatibility in natural populations of freshwater sponges. *J. Exp. Zool.* 237:241–255.

Müller, W.E.G., Gamulin, V., Rinkevich, B., Spreitzer, I., Weinblum, D. and Schroder, H.C. (1994a) Ubiquitin and ubiquination in cells from the marine sponge *Geodia cydonium. Biol. Chem. Hoppe-Seyer* 375:53–60.

Müller, W.E.G., Müller, I.M. and Gamulin, V. (1994b) On the monophyletic evolution of the metazoa. *Brazilian J. Med. Biol. Res.* 27:2083–2096.

Nakao, Y., Matsunaga, S. and Fusetani, N. (1993a) Toxadocials B, C and toxadocicc acid a – thrombin-inhibitory aliphatic tetrasulfates from the marine sponge, *Toxadocia cylindrica. Tetrahedron* 49:11183–11188.

Nakao, Y., Matsunaga, S. and Fusetani, N. (1993b) Toxadocial A – a novel thrombin inhibitor from the marine sponge *Toxadocia cylindrica. Tetrahedron Lett.* 34:1511–1514.

Pettit, G.R., Tan, R., Gao, F., Williams, M.D., Doubek, D.L., Boyd, M.R., Schmidt, J.M., Chapuis, J.C., Hamel, Bai, E., Hooper, J.N.A. and Tackett, L.P. (1993) Isolation and structure of halistatin-1 from the eastern Indian ocean marine sponge *Phakellia carteri. J. Org. Chem.* 58: 2538–2543.

Pettit, G.R., Gao, F., Cerny, R.L., Doubek, D.L., Tackett, L.P., Schmidt, J.M. and Chapuis, J.C. (1994) Antineoplastic agents 278. Isolation and structure of axinastatins 2 and 3 from Western Caroline Island marine sponge. *J. Med. Chem.* 37:1165–1168.

Pfeifer, K., Schroder, H.C., Rinkevich, B., Uhlenbruck, G., Hanisch, F.G., Kurelec, B., Scholz, P. and Muller, W.E.G. (1992) Immunological and biological identification of tumor necrosis-like factor in sponges: endotoxin that mediates necrosis formation in xenografts. *Cytokine* 4: 161–169.

Pfeifer, K., Frank, W., Schroder, H.C., Gamulin, V., Rinkevich, B., Batel, R., Muller, I.M. and Muller, W.E.G. (1993) Cloning of the polyubiquitin cDNA from the marine sponge *Geodia cydonium* and its preferential expression during reaggregation of cells. *J. Cell Sci.* 106:545–554.

Proksch, P. (1994) Defensive roles for secondary metabolites from marine sponges and sponge-feeding Nudibranchs. *Toxicon* 32:639–655.

Spillmann, D., Hard, K., Thomasoates, J., Vliegenthart, J.F.G., Misevic, G., Burger, M.M. and Finne, J. (1993) Characterization of a novel pyruvylated carbohydrate unit implicated in the cell aggregation of the marine sponge *Microciona prolifera. J. Biol. Chem.* 268: 13378–13387.

Sullivan, B., Djura, P., McIntyre, D.E. and Faulkner, D.J. (1981) Antimicrobial constituents of the sponge *Siphonodictyon coralliphagum. Tetrahedron* 37:979–982.

Takei, M., Umeyama, A., Shoji, N., Arihara, S. and Endo, K. (1993) Mechanism of inhibition of IgE-dependent histamine release from rat mast cells by xestobergsterol A from the Okinawan marine sponge *Xestospongia bergquistia. Experientia* 49:145–149.

Teeyapant, R., Woerdenbag, H.J., Kreis, P., Hacker, J., Wrav, V., Witte, L. and Proksch, P. (1993a) Antibiotic and cytotoxic activity of brominated compounds from the marine sponge *Verongia aerophora. Z. Naturforsch.* 48:939–945.

Teeyapant, R., Kreis, P., Wray, V., Witte, L. and Proksch, P. (1993b) Brominated secondary compounds from the marine sponge *Verongia aerophoba* and the sponge feeding gastropod *Tylodina pervesa. Z. Naturforsch.* 48c:939–945.

Thompson, J.E. (1985) Exudation of biologically-active metabolites in the sponge *Aplysina fistularis.* I. Biological evidence. *Marine Biol.* 88:23–26.

van de Vyver, G. (1981) Organisms without special circulatory systems. *In:* N.A. Ratcliffe and A.F. Rowley (eds.): *Invertebrate Blood Cells, Vol. 1,* Academic Press, New York, pp 19–32.

Vasta, G.R. (1991) The multiple biological roles of invertebrate lectins: their participation in nonself recognition mechanisms. *In:* W.G. Warr and N. Cohen (eds): *Phylogenesis of Immune Function,* CRC Press, Boca Raton, pp 73–202.

Wissenfels, N. (1976) Bau und Funktion des Süsswasserschwamms *Ephydatia fluviatilis.* III. Nahrungsaufnahme, Verdauung und Defäkation. *Zoomorph.* 85:73–88.

Willenz, P. (1980) Kinetic and morphological aspects of particle ingestion by the freshwater sponge *Ephydatia fluviatilis. In:* D.C. Smith and Y. Tiffon (eds): *Nutrition in the Lower Metazoa,* Pergamon Press, Oxford, pp 163–178.

Wilson, E.V. (1907) On some phenomena of coalescence and regeneration in sponges. *J. Exp. Zool.* 5:245–258.

Diblastic animals

Coelenterata

The radial symmetry of the blind-sac basic body construction and the bilayered body wall are taxonomic justification for uniting two phyla, the *Coelenterata (Cnidaria)* and the *Ctenophora,* within the superassemblage names *Diblastica* and *Radiata.* The cnidarians (more than 9000 species) consist of three classes, the *Hydrozoa* (familiar hydras, hydroids), the *Scyphozoa* (jellyfish), and the *Anthozoa* (sea anemones, corals). The biradiate comb jellies and sea walnuts form the sister group of cnidarians, the phylum *Ctenophora* (50 species) (Barnes, 1987).

Two basic features join the *Coelenterata* with all metazoans, the internal digestive space (called gastrovascular cavity) and the presence of the mouth. The bilayered body wall is composed of the outer epithelium (ectoderm) and inner digestive lining (endoderm), which are separated by a jelly-like mucoid layer, the mesoglea, with or without the presence of freely wandering amoeboid cells of ecto- and endodermal origin.

Coelenterates are endowed by a number of more progressive features in comparison with previous sponges: their cells are apparently more specialized, and for the first time in evolution, the morphofunctionally specialized true tissues appear, including the nervous system. The gastrovascular cavity can be transformed in more advanced species into specialized systems mimicking effective respiratory, excretory, and vascular functions (Kampmeier, 1969).

From the defense point of view, nothing is known about immune reactions of the *Ctenophora* which are delicate and fragile, exclusively marine, animals. Most of the research has been devoted to the Hydrozoans. With the exception of the classical works dealing with phagocytosis (Metchnikoff, 1892), most of the research focused on the immunological mechanisms of these creatures and was done only recently. Significant attention was paid to the external defensive reactions, oriented towards rather large animals, but these reactions are not immunological in the true meaning. The coelenterates are equipped with stinging cells (cnidocytes, nematocytes), serving in defense and also in prey capture. The secretory products of these specialized cells have extremely strong and potentially even deadly effects (Burnett et al., 1986; Halstead, 1987; Dimarzo et al., 1993; Endean et al., 1993; Salleo et al., 1993; Tardent, 1995).

Humoral immunity

In contrast to the rather sophisticated activities of cellular immunity of Coelenterates, only a limited attention has been paid to the studies of the humoral branch of their defense reactions. Mucus secreted by epidermal gland cells is probably involved in forming a nonspecific mechanico-chemical barrier. Mucus released by *Anthopleura elegantissima* has a pH of 5.9 and is capable of lytic action against some bacteria. *Micrococcus lysodeicticus* is the most sensitive bacteria to this mucus, thus the mucus properties resemble actions of lysozyme (Phillips, 1963). Lubbock in 1979 suggested that anemone mucus may be species-specific and may represent the key factor needed for immune recognition (Lubbock, 1979). The mucus specificity could be exerted in various extents, however, while mucus from congeneric *A. xanthogrammica* failed to elicit the discharge of tentacular nematocytes in *A. elegantissima*, the latter's mucus induced the discharge in *A. xanthogrammica* (Ertman and Davenport, 1981). Production of components with antimicrobial activities was reported in some individuals belonging to *Alcyonacea, Scleratinia, Zooanthidea,* and *Siphonophora* (Burkholder, 1973; Burkholder and Burkholder, 1958). A substance with an antileukemia activity has been isolated from the soft coral *Lemnalia africana* (Jurek and Scheuer, 1993).

Another substance, a cytolysin called helianthin, has been isolated from the sea anemone, *Stichodactyla helianthis*. It is a basic polypeptide with a specific affinity for membrane sphingomyelin similar to other peptides found in nematocyte toxins of various sea anemones. Up to now, at least 16 various membrane damaging toxins have been isolated. On the contrary, an anemone *Metridium senile*, the most common of all large anemones, contains anacidic cytolytic protein of about 80 kDa, closely similar to the thiol-activated cytolysins of bacterial origin (Bernheimer, 1996).

Unfortunately, the agglutinins, important substances in the immune recognition processes of almost all animal phyla, have been studied only in gorgonacean corals. *D*-mannose-specific lectin inhibiting mRNA transport through nuclear membrane pores was isolated in *Gererdia savaglia* (Kljajic et al., 1987).

Neither antigen-inducible humoral defense factors comparable to vertebrate antibodies bearing the immunoglobulin character nor the non-specific inducible factors known in various invertebrate phyla have been found in coelenterates. The only exception is the observation by Phillips and Yardley (1960) in *A. elegantissima* of an inducible substance with a high affinity to the antigen used for immunization.

Cellular immunity

The immune phenomena characteristics
In addition to the freely moving phagocyte cells, three types of epidermal cells are significant in the view of cnidarian defense: the cnidocytes, the mucus-secreting cells, and the interstitial cells (I-cells) representing a stem cell type giving rise to germinal cells as well as to all other cells including the immune ones. There are numerous types of cnidocytes in some coelenterate species which have functions that are highly specialized, either for capturing of a prey, or for aggression towards members of other clones (Bigger, 1980, 1982). These highly discriminative functions are attributed to the immune recognition of surface specific structures, probably the same recognition by which the alloimmune reaction is triggered (Hildemann et al., 1975). The acrorhagial reaction, i.e. the discharging of acrorhagial cnidocytes, used by some sea anemones against the individuals of different clones, is an example of a specific immune reaction comparable analogically to the T-cell immunity of vertebrates in the sense that the killing is stronger towards allogeneic than xenogeneic relations (Francis, 1973; Bigger, 1980). There is, moreover, an interesting recognition functional dichotomy. Whereas the tentacular nematocytes react to a wide range of diverse animals, the acrorhagial cnidocytes are more specific: they respond by their discharging only to certain allogeneic or xenogeneic corals (Lubbock, 1980).

Like in all invertebrates, coelenterates defend their internal milieu primarily by phagocytosis. However, there exists a new twist: a large number of coelenterates use protozoa as intracellular symbionts. Some of these cytobionts are heterotrophic parasites and can be lethal for the host. On the other hand, others might be beneficial. Endosymbiosys in *Hydra* is a classical model for the study of the evolutionary aspects of immunity. Certain species of *Chlorella* occur naturally as cytobionts in *Hydra viridissima*. The algae evoke phagocytosis by digestive cells and are sequestered in individual phagosomes that migrate to the base of the cells where they resist fusion with lysosomes. *Hydra* controls the algal population by digesting excess algae. This type of symbiosis is mutually beneficial (Muscatine, 1989; Muscatine and McNeil, 1989).

The presence of living algae inside phagocytic cells is a unique characteristic. Besides these cells, the other main cell type involved in phagocytosis is the free amoebocyte (mesogleal cell, granulocyte). Based on a very limited number of studies, there are basic differences in phagocytosis among individual classes of Coelenterates (Sparks, 1972; Tokin and Yericheva, 1961).

In contrast to the members of triblastic taxa, the encapsulation processes in coelenterates have been described only rarely. Only encapsulation of metacercariae of a trematode *Plagioporous* species invading corals, *Porites* (*Scleratinia*), or some invasive algae in *Gorgonacea* has been observed (Cheney, 1975; Morse et al., 1977).

A unique feature of Coelenterates is their phenomenal ability of regeneration, which was described back in 1744 (Trembley, 1744). The speed and quality of regeneration reduced the probability of secondary infection and thus a need for development of any complex immune mechanisms (Sparks, 1972). In anthozoans and scyphozoans the cooperation of various epidermal and gastrodermal cell types together with interstitial and amoeboid cells (Tardent, 1980) has been observed. Moreover, first in evolution the participation of neurotransmitters controlling the regeneration processes has been proved (Grimmelihuijzen et al., 1992).

By 1991, at least 72 species of hydrozoans and anthozoans have been studied (Leddy and Green, 1991), but the results of the studies of histocompatibility recognition in Coelenterates are rather confusing. In some species allogeneic or xenogeneic incompatibilities should be ascribed to the mechanisms of the recognition of self, identical or similar (or blindness to non-self), like in *Gorgonia* which shows non-acceptance of foreign tissue as rejection (Theodor, 1970). In the hydrozoans, most of the studies on allogeneic interactions were performed on *Hydra* (Znidaric, 1981; Shimizu and Sawada, 1987) and *Hydractinia* species (Hauenschild, 1954; Ivker, 1972; Lange et al., 1989).

The intensity of the histocompatibility reaction in *Hydrozoa* and *Anthozoa* differs based on stage, e.g., *Podocaryne carnea* tolerates allografts at the early stage of development, but rejects them as an adult (Tardent and Buhrer, 1982). Allografts are often tolerated, leading to the development of a chimera (Campbell and Bibb, 1970). A colonial hydroid *Hydractinia echinata* was studied in more detail. It was found that one (and probably more) locus is responsible for fusion control (Hauenschild, 1954, 1956). It is of interest to note that this model was used for first demonstration of a genetically coded defect in autoreactivity in invertebrates. Some individuals formed hyperplastic stolons during genetical clonings and these stolons were reacting against self tissues (Buss et al., 1985).

Coelenterates developed several other reactions which belong to the histocompatibility recognition, but are used for external defense instead of a more common defense of the internal millieu. Some of these reactions are exocoelenteral digestion, barrier formation (Hildemann et al., 1975, 1977), contact autolysis (Muller et al., 1984) and development of specialized sweeper tentacles (Hidaka and Yamazato, 1984). High frequency of repeated contacts between genetically different individuals is probably the main cause of the development of the special type of transplantation memory. *A. elegantissima* is able to extremely shorten the time interval between aggressive attacks after repeated contacts (from 15 min to 30 s) (Leddy and Green, 1991).

Histocompatibility reaction in *Montipora verrucosa* is the best studied allogeneic reactivity in all invertebrates. The allograft rejection is accompanied with highly discriminative temperature-dependent cytotoxic reaction and short-lived immunological memory, controlled by polymorphic H-gene complex (Hildemann et al., 1975, 1977, 1980; Raison et al., 1976). The gorgonial coral *Swiftia exserta* fulfills all basic criteria of cytotoxicity, specificity and memory. All autografts are fused and all allografts are rejected. The secondary allografts are rejected in an elevated rate (from 7–9 days to 3–4 days) (Salter-Cid and Bigger, 1991).

A complex allorecognition system was also found in a reef-building coral, *Srylophora pistillata* (Chadwick-Furman and Rinkevich, 1994). The authors assayed branch pair combinations among 11 colonies for 24 months and found that different allogeneic combinations exhibited either unilateral rejection, or an array of other incompatible reaction. A self/non-self histocompatibility recognition was also demonstrated in a solitary reef coral *Fungia scutaria* (Jokiel and Bigger, 1993), the response being analogous to that observed in colonial corals (Hildemann et al., 1975). Similar observations were achieved when *Millepora dichotoma* (Frank and Rinkevich, 1994) or *Acropora hemprichi* (Rinkevich et al., 1994) was tested. For an excellent, more complex and most recent review of cnidarian histocompatibility, see that by Leddy and Green (1991).

Conclusions

Despite the fact that the members of the phylum *Coelenterata* reached only a primitive stage in the evolutionary pathway, they have developed well differentiated true tissues and first distinct organs within the metazoans. They are extensively able to distinguish self and non-self and efficiently react against a plethora of foreign challenges. Within the coelenterate phylum a shift from non-specific immune reaction in *Hydrozoa* to sharply discriminative immune response in *Anthozoa* could be traced. The anthozoans do perform the more cooperative, genetically controlled response accompanied by a short immunological memory of which the crucial effector cells remain the amoebocytes. The molecules and the cell surface receptors, and effector molecules participating in both recognition and cytotoxic response, remain to be discovered.

Humoral immunity is only represented by lectins bearing antibacterial activities and mainly by various toxic substances with wide, more or less non-specific effects. The coelenterates have developed some bizzare defense responses, some of them behavioral rather than immune. Evolutionary constriction in morphofunctional patterns of coelenterates has been compensated by a highly-developed capability of regeneration never seen among remaining eumetazoan assemblages.

References

Barnes, R.D. (1987) *Invertebrate Zoology* (5th ed.) Saunders Coll. Publ., Philadelphia.
Bernheimer, A.W. (1996) Some aspects of the history of membrane-damaging toxins. *Med. Microbiol. Immunol.* 185:59–63.
Bigger, C.H. (1980) Interspecific and intraspecific acrorhagial aggressive behavior among sea anemones, a recognition of self and non-self. *Biol. Bull.* 159:117–122.
Bigger, C.H. (1982) The cellular basis of the aggressive acrorhagial response of sea anemones. *J. Morphol.* 173:259–278.
Burkholder, P.R. (1973) The ecology of marine antibiotics and coral reefs. *In:* O.A. Jones and R. Endean (eds): *Biology and Geology of Coral Reef,* Academic Press, New York, pp 117–182.
Burkholder, P.R. and Burkholder, L.M. (1958) Antimicrobial activity of horny corals. *Science* 127:1174–1176.
Burnett, J.W., Calton, G.J. and Burnett, H.W. (1986) Jellyfish envenomation syndromes. *J. Am. Acad. Dermatol.* 14:100–106.
Buss, L.W., Moore, J.L. and Green, D.R. (1985) Autoreactivity and self tolerance in an invertebrate. *Nature* 313:400–402.
Campbell, R.D. and Bibb, C. (1970) Transplantation in coelenterates. *Transpl. Proc.* 2:202–212.
Chadwick-Furman, N. and Rinkevich, B. (1994) A complex allorecognition system in a reef-building coral: delayed responses, reversals and nontransitive hierarchies. *Coral Reefs* 13:57–63.
Cheney, D.P. (1975) Hard tissue tumors of scleratinian corals. *In:* W.H. Hildemann and A.A. Benedict (eds): *Immunologic Phylogeny,* Plenum Press, New York, pp 77–87.
Dimarzo, V., Depetrocellis, L., Gianfrani, C. and Cimino, G. (1993) Biosynthesis, structure and biological activity of hydroxyeicosatetraenoic acids in *Hydra vulgaris. Biochem. J.* 295:23–29.
Endean, R., Monks, S.A. and Cameron, A.M. (1993) Toxins from the box jellyfish *Chironex fleckeri. Toxicon* 31:397–410.
Ertman, S.C. and Davenport, D. (1981) Tentacular nematocyte discharge and "self-recognition" in *Anthopleura elegantissima* Brant. *Biol. Bull.* 161:366–370.
Francis, L. (1973) Intraspecific aggression and its effect on the distribution of *Anthopleura elegantissima* and some related sea anemones. *Biol. Bull.* 144:73–92.
Frank, U. and Rinkevich, B. (1994) Nontransitive patterns of historecognition in the Red Sea hydrocoral *Millepora dichotoma. Marine Biol.* 118:723–729.
Grimmelihuijzen, C.J.P., Carstensen, K., Darmer, D., Moosler, A., Nothacker, H.P., Reinscheid, R.K., Schmutzler, C. and Vollert, H. (1992) Coelenterate neuropeptides: structure, action and biosynthesis. *Amer. Zool.* 32:1–12.
Halstead, B.W. (1987) Coelenterate (cnidarian) stings and wounds. *Clin. Dermatol.* 5:8–13.
Hauenschild, V.C. (1954) Genetische und entwicklungphysiologische Untersuchungen über Intersexualität und Gewebeverträglichkeit bei *Hydractinea echinata. Flem. Wilhelm Roux Arch.* 147:1–141.
Hauenschild, V.C. (1956) Über die Vererbung einer Gewebeverträglichkeitseigenschaft bei dem Hydriodpolypen *Hydractinea echinata. Z. Naturforsch.* 11:132–143.
Hidaka, M. and Yamazato, K. (1984) Intraspecific interactions in a scleractinian coral, *Galaxea fascicularis*: induced formation of sweeper tentacles. *Coral Reefs* 3:77–85.
Hildemann, W.H., Linthicum, D.S. and Vann, D.C. (1975) Transplantation and immuno-incompatibility reactions among reef-building corals. *Immunogenetics* 2:269–284.
Hildemann, W.H., Raison, R.L., Cheung, G., Hull, C.J., Akala, L. and Okamoto, J. (1977) Immunological specificity and memory in a scleractinian coral. *Nature* 270:219–223.
Hildemann, W.H., Jokiel, P.L., Bigger, C.H. and Johnston, I.S. (1980) Allogenic polymorphism and alloimmune memory in the coral, *Montipora verrucosa. Transplantation* 30:297–301.
Ivker, F.B. (1972) A hierarchy of histocompatibility in *Hydractinia echinata. Biol. Bull.* 143:162–174.
Jokiel, P.L. and Bigger, C.H. (1994) Aspects of histocompatibility and regeneration in the solitary reef coral *Fungia scutaria. Biol. Bull.* 186:72–80.

Jurek, J. and Scheuer, P.J. (1993) Sesquiterpenoids and norsesquiterpenoids from the soft coral *Lemnalia africana. J. Nat. Products* 56:508–513.

Kampmeier, O.F. (1969) *Evolution and Comparative Morphology of the Lymphatic System.* C.C. Thomas Publ., Springfield.

Kljajic, A., Schröder, H.C., Rottmann, M., Cuperlovic, M., Movsesian, M., Uhlenbruck, G., Gasic, N., Zahn, R.K. and Müller, W.E.G. (1987) A D-mannose-specific lectin from *Gerardoa savaglia*, that inhibits nucleocytoplasmic transport of mRNA. *Eur. J. Biochem.* 169:97–104.

Lange, R., Plickert, G. and Müller, W.A. (1989) Histocompatibility in a low invertebrate. *Hydractinia echinata*: analysis of the mechanism of rejection. *J. Exp. Zool.* 249:284–292.

Leddy, S.V. and Green, D.R. (1991) Historecognition in the *Cnidaria. In:* W.G. Warr and N. Cohen (eds): *Phylogenesis of Immune Functions,* CRC Press, Boca Raton, pp 103–116.

Lubbock, R. (1979) Mucus antigenicity in sea anemones and corals. *Hydrobiol.* 66:3–6.

Lubbock, R. (1980) Clone-specific cellular recognition in a sea anemone. *Proc. Nat. Acad. Sci. U.S.A.* 77:6667–6669.

Metchnikoff, E.E. (1892) *Leons sur la patologie Comparee de l'Inflammation.* Masson, Paris.

Morse, D.E., Morse, A.N.C. and Duncan, H. (1977) Algal "tumors" in the Caribean sea-fan, *Gorgonia ventalina. In:* D.L. Taylor (ed.): *Proceedings of the Third International Coral Reef Symposium, Vol. I,* University of Miami, Miami, pp 623–629.

Müller, W.E.G., Müller, I., Zahn, R.K. and Maidhof, A. (1984) Intraspecific recognition system in scleractinian corals: morphological and cytochemical description of the autolysis mechanism. *J. Histochem. Cytochem.* 32:285–288.

Muscatine, L. (1989) Adventures in symbiosis. *Amer. Zool.* 29:1203–1208.

Muscatine, L. and McNeil, P.L. (1989) Endosymbiosis in *Hydra* and the evolution of internal defense system. *Amer. Zool.* 29:371–286.

Phillips, J.H. (1963) Immune mechanisms in the phylum Coelenterate. *In:* E.C. Dougherty, Z.N. Brown, E.D. Hanson and W.D. Hartman (eds): *The Lower Metazoa,* University of California Press, Berkeley, pp 425–431.

Phillips, J.H. and Yardley, J.B. (1960) Detection in invertebrates of inducible, reactive materials resembling antibody. *Nature* 188: 728-730.

Raison, R.L., Hull, C.J. and Hildemann, W.H. (1976) Allogeneic graft rejection in *Montipora verrucosa*, a reef-building coral. *In:* R.K. Wright and E.L. Cooper (eds): *Phylogeny of Thymus and Bone Marrow/Bursa Cells,* Elsevier/North Holland, Amsterdam, pp 3–8.

Rinkevich, B., Frank, U., Bak, R.P.M. and Müller, W.E.G. (1994) Alloimmune responses between *Acropora hemprichi* conspecifics: nontransitive patterns of overgrowth and delayed cytotoxicity. *Marine Biol.* 118:731–737.

Salleo, A., Santoro, G. and Barra, P. (1993) Spread of experimentally induced discharge of the nematocytes in acontia of *Calliastis parasitica. Comp. Biochem. Physiol. [A]* 104:565–574.

Salter-Cid, L. and Bigger, C.H. (1991) Alloimmunity in the Gorgonial coral *Swiftia exserta. Biol. Bull.* 181:127–134.

Shimizu, H. and Sawada, Y. (1987) Transplantation phenomena in *Hydra*: cooperation of position-dependent and structure-dependent factors determining the transplantation results. *Dev. Biol.* 122:113–119.

Sparks, A.K. (1972) *Invertebrate Pathology.* Academic Press, New York.

Tardent, P. (1995) The cnidarian cnidocyte, a high-tech cellular weaponry. *Bioessays,* 17:351–362.

Tardent, P. and Buhrer, M. (1982) Intraspecific tissue incompatibilities in the metagenetical *Podocaryne carnea* M. Sars (Cnidaria: Hydrozoa). *In:* M.M. Burger and R. Weber (eds): *Embryonic Development. Part B. Cellular Aspects,* Alan R. Liss, New York, pp 295–315.

Tardent, P. and Tardent, R. (1980) *Developmental and Cellular Biology of Coelenterates.* Elsevier/North-Holland Biomedical Press, Amsterdam.

Theodor, J.L. (1970) Distinction between "self" and "non-self" in lower invertebrates. *Nature* 227:690–692.

Tokin, B.P. and Yericheva, F.N. (1961) Phagocytal reaction in the course of regeneration and somatic embryogenesis in lower coelenterates. *Tr. Murm. Morsk. Inst.* 3:182–192.

Trembley, A. (1744) *Memoires pour servir a l'historie d'un genre de polypes d'eau douce a bras en ferme de cornes.* Leyden.

Znidaric, D. (1981) Regeneration of autografts, isografts and xenografts of *Hydra. Period. Biol.* 83:295–299.

Protostomes

Annelida

From general points of environmental and agroeconomic importance, the annelids, and particularly the oligochaetes, have attracted scientific attention for more than a hundred years. Charles Darwin (1881) summarized their natural role concisely: "It may be doubted whether there are many other animals which have played so important a part in the history of the world, as have these lowly organized creatures." Since the early 1960s a group of researchers headed by E.L. Cooper have investigated the fundamental immune mechanisms of earthworms. Plenty of data has been accumulated up to the present, which has converted the earthworms into a general experimental model of invertebrate immunity, "the mice" of invertebrates. The most recent review of earthworm immunity is by Cooper (1996).

Whereas platyhelminthes appear to be the most primitive triploblastic animals, the annelids are considered to be the ancestral key assemblage in which important evolutionary novelties, the metamery (segmentation) and secondary body cavity (coelom) had emerged. From this point of view, they represent virtually a new, progressive monophyletic group in the scale of increasing structural complexity of metazoans. The emergence of new structures and organs brought about better dynamics of locomotion and new, more effective functional and immune properties facilitating their environmental adaptibility (Clark, 1964). Due to this, many annelid lineages have diversified, comprising between 8700–9000 estimated species classified into two main superclasses, the *Aclitellata,* in which marine polychaetes *(Polychaeta)* form the most important group, and the *Clitellata* consisting of well-known earthworms *(Oligochaeta)* and leeches *(Hirudinea)* (Barnes, 1987, 1989; Brinkhurst, 1991). More details about phylogeny, phyletic relationships, and classification of annelid subtaxa are referred to in Síma (1994a, b).

From three classes forming the *Annelida* phylum, only *Oligochaeta,* and in fact only earthworms, are discussed. This limitation is primarily caused by the fact that earthworms constitute inexpensive and relatively simple experimental model which has been extensively studied for decades. In addition, there is a lack of information about other annelids. Readers seeking more detailed data about immunology and physiology of annelids should see Větvička et al. (1994) and Tučková and Bilej (1996). The gene-

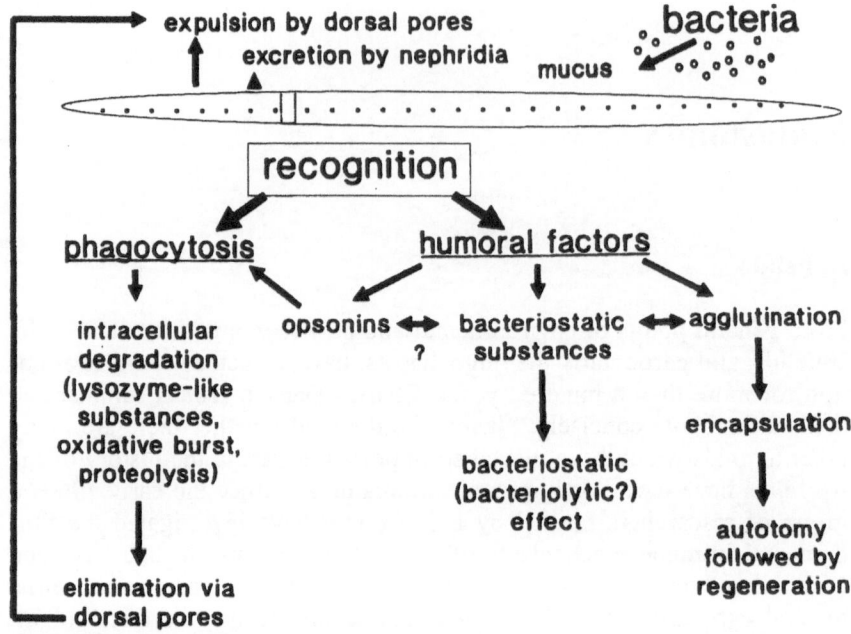

Figure 1. General armamentarium of natural resistance of the earthworm. (From Tučková and
Bilej, 1996, with permission.)

ral scheme showing the main resources of earthworm defense are shown in
Figure 1.

The evolutionary significance of metamery and the coelom

The annelids posses all three major advanced structural features over previous groupings of metazoans: the coelom, the closed vascular system, and
the metameric arrangement of their bodies. The coelom has two progressive features: first it represents an effective hydraulic device for movement;
secondly, it provides a space for development of more complex organs. The
metamerism in the course of phylogeny became a device for increased
regional differentiation. Since the time of appearance of both coelom and
metamery, all major phyla, the arthropods and chordates have utilized their
evolutionary advantages. "The duplication and subsequent divergence of
genes allows genomes to increase in complexity, which in turn is probably
necessary for increases in the complexity of whole organisms," wrote
Graham (1995). This means that the redundancy in segmentaion is reflexed
in the genome in which the processes of tandem duplications of predestin-
ed genes have become later in phylogeny a significant part of the emergence of an advanced stage of complexity (Ohno, 1970). Similarly, the

same mechanisms led to the appearance of various defense molecules of an immunoglobulin superfamily (Hill et al., 1966; Singer and Doolitle, 1966). Moreover, the evolutionary advantage of metamery is capability for regional differentiation, in other words, capability not only to produce identical segments but also to change duplicated parts in ways that would be advantageous for a whole organism. More complex and cooperative immune reactions that we know in annelids could evolve only in such coelomatic segmented animals in which the regional specialization of their body patterns has been the *conditio sine qua non* for functional specialization.

The cells

A relatively high number of types of free-wandering cells can be found both in the coelomic fluid and inside the vascular system. From the point of a basic orientation, the free cells involved in the immune processes (such as recognition, phagocytosis, encapsulation, and histocompatibility reaction) are various categories of coelomocytes which can be generally divided into main categories like amoebocytes, eleocytes, erythrocytes, and hemocytes in polychaetes, similarly, amoebocytes, eleocytes, and hemocytes in oligochaetes, and amoebocytes and free chloragogen cells in hirudineans (see Tab. 1). For the more detailed survey and further reading see Síma (1994c).

The origin

Only exceptionally have distinct cytopoietic structures been described in annelids. In polychaetes, the distinct "lymph glands" and paired structures located in coelomic peritoneum have been described. They arise from specialized parts of coelomic or blood vessel epithelium, or from the distinc-

Table 1. A comparative survey of the main annelid cell types involved in the defense processes (the most frequent synonyms are in parentheses)

Category	*Polychaeta*	*Oligochaeta*	*Hirudinea*
Amoebocyte (granulocyte)	+	+	+
Eleocyte (trephocyte)	+	+ (chloragocyte)	+ (chloragogen cell)
Erythrocyte (hemocyte)*	+	–	–
Hemocyte (blood cell)	+	+	

* misleading, exclusively in coelomic cavity.

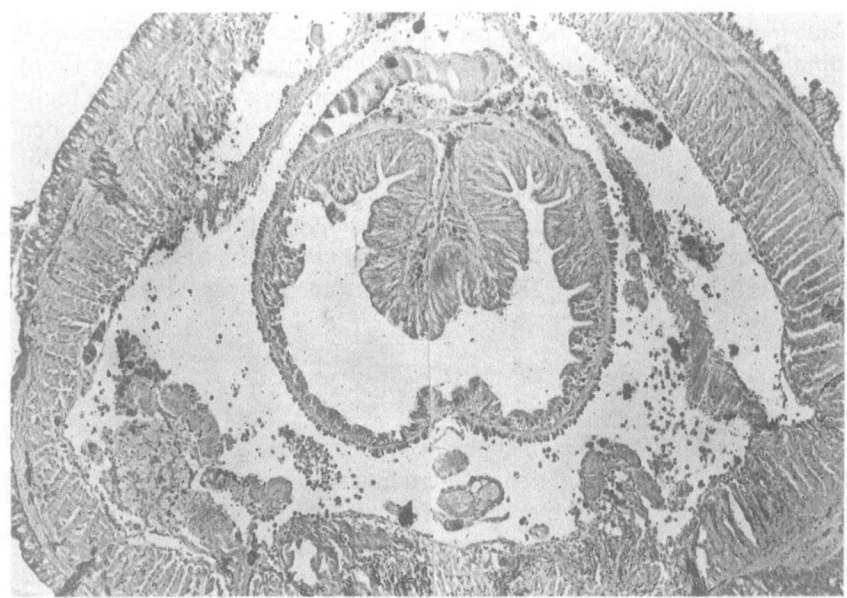

Figure 2. Localization of I^{125}-labeled HEMA particles through the coelom and typhlosole of *Lumbricus terrestris*. Autoradiography (magnification 200×).

tive structures described as "lymph glands" (Dales and Dixon, 1981). Similar structures or hemogenic organs called "lymphoid organs," "blood glands," or "blood follicles" have been described in some genera of oligochaetes (Stephenson, 1924; Friedman and Weiss, 1982). Later on these structures were considered rather phagocytic. In the prevailing number of annelid worms studied the origin of free coelomic cells is determined from the peritoneal and gut wall tissues, coelomic splanchnic and somatic epithelia, and epithelia covering blood vessels (for review see Cooper and Stein, 1981; Dales and Dixon, 1981; Sawyer and Fitzgerald, 1981). In earthworms the antigenic material is preferentially captured and usually processed from these locations, and also inside the typhlosole (see Figs 2, 3).

Humoral immunity

An important part of the annelids' defense reactions relies on proteolytic enzymes. These enzymes are used in proteolytic degradation of internalized material (Valembois et al., 1973) including vertebrate proteins (Tučková et al., 1986). These enzymes resemble chymotrypsin. The proteolytic activity significantly differs among earthworm species. Nine different serine proteinases were isolated from *Sabellaria alveolata* (Peaucellier, 1983) and three from *Eisenia foetida* (Roch et al., 1991). Further character-

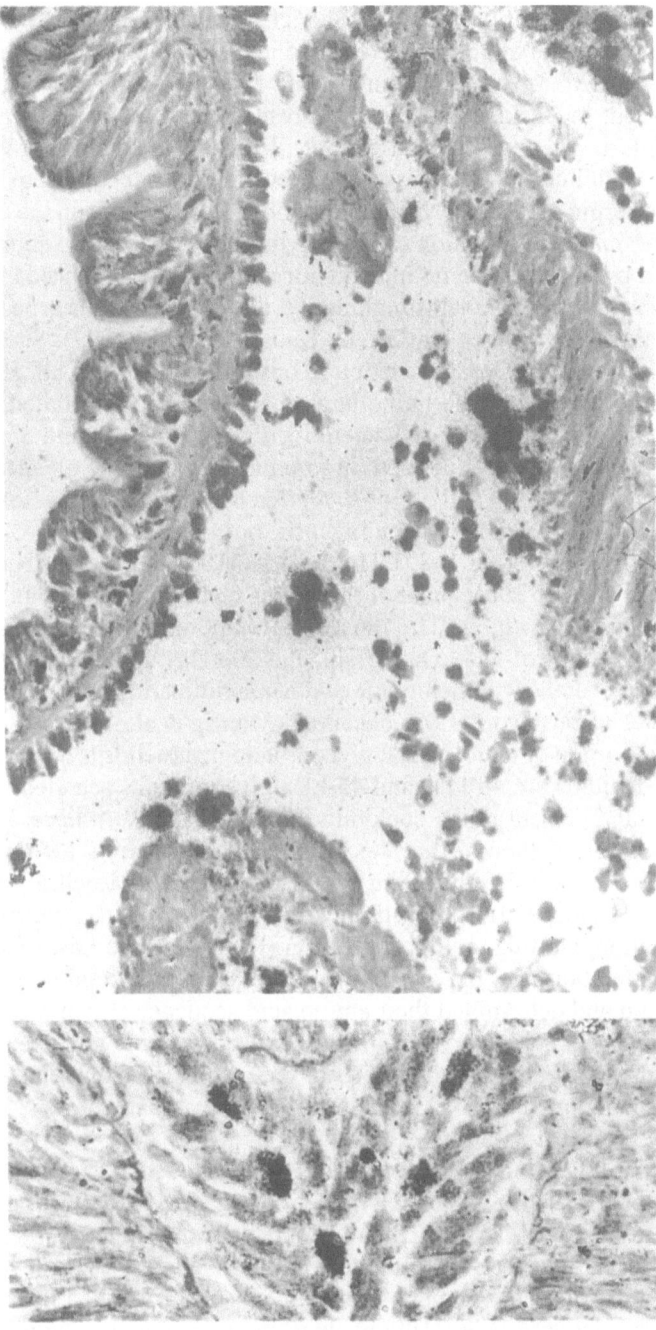

Figure 3. Details of the localization of I^{125}-labeled HEMA particles through the coelom and typhlosole of *Lumbricus terrestris*. Autoradiography (magnification 400×).

ization showed their trypsin-like and chymotrypsin-like characteristics. Very strong trypsin inhibitors were found in the coelomic fluid of the earthworm *Lumbricus terrestris* (Voburka et al., 1992). Two inhibitors with molecular weight of 42 kDa and 20 kDA, respectively, are present in several forms differing in their isoelectric points. Based on the molecular weight of the larger inhibitor, one can assume an evolutionary relationship with the family of serpin inhibitors. No similar inhibitors were found in the coelomic fluid of *E. foetida*. Formation of proteolytic substances is also significantly increased by intracoelomatic injection of antigen. The enhanced activity is seen within 24 h and remains high until 8 days later. Various aspects of the proteolytic activity were analyzed by Kauschke et al. (1997). Some of the hemolysis has an additional function as opsonins (Sinkora et al., 1995).

The original tests of the humoral antibacterial response failed to show any significant inhibition of bacterial growth. The apparent setback of these studies was the use of bacteria foreign to the biotope of earthworms. When the antibacterial activity of *E. foetida* has been tested against strains originating in the same surroundings, six out of 23 strains have been inhibited (Valembois et al., 1982). This bacteriostatic activity was been found to be thermolabile. Subsequent observations revealed 11 fractions with differing molecular weights. The most significant antibacterial activity was found in fractions of 175 kDa, 45, 40, and 20 kDa (Vallier et al., 1985). A strong crossreactivity of hemolytic and hemaglutinating proteins in different species of earthworms was observed (Mohrig et al., 1996).

Another group of antibacterial glycoprotein in annelids is represented by fetidins. Fetidins are 40-kDa and 45-kDa glycoproteins secreted by chloragocytes and present in the coelomic fluid of *E. foetida andrei*. Immunoelectrofocusation showed that the 45-kDa glycoprotein is monomorphic, while at least four isoforms of the 40-kDa species exist (Roch, 1979). They posses a broad spectrum of activities including hemolysis, bacteriolysis, agglutination, clotting, and opsonization (for review see Lassagues et al., 1997). Valembois and his group successfully isolated fetidins from coelomic fluid and determined their amino acid sequence (Roch et al., 1989). In subsequent experiments the same group used screening of an expression cDNA library from earthworm total tissue and isolated and characterized cDNA encoding the 40-kDa fetidin. The recombinant protein inhibited bacterial growth (Lassagues et al., 1997).

Bacterial agglutinins of *L. terrestris* were studied by Cooper's group (Cooper et al., 1974; Stein et al., 1986; Stein et al., 1990). Several agglutinins were observed, some of them inducible after specific (Stein et al., 1986) or even nonspecific stimulation (Kauschke and Mohrig, 1987a).

For summary of the hemagglutinating systems described in individual annelid species, see Bilej (1994b). In *E. foetida*, four hemagglutinins of MW 11.5, 20, 30, and 40 kDa have been isolated (Roch et al., 1984). The hemagglutination titers widely differ depending on the type of erythrocytes used in the assay. In *L. terrestris*, hemagglutination is inhibited by carbohydrates (Stein and Cooper, 1983). The cells involved in production of these factors are amoebocytes (Mohrig and Kauschke, 1984), which is different from the situation found in *Eisenia*.

E. foetida has been traditionally used as a model of the hemolytic systems in annelids. The first thermolabile factor produced by chloragocytes has been described as early as 1968 (Du Pasquier and Duprat, 1968). Roch's group isolated four different proteins of pI between 5.9 and 6.3. All earthworms possess either two or three isoforms (Roch, 1979). There are interesting differences between the European and American population of earthworms regarding the allelic forms coding these proteins. One-third of the American earthworms expressed a fourth allele which is never found in the European creatures. Valembois and Roch hypothesized that this allele was originally a rare allele in the ancestral European earthworms, which was more expressed after migration to America (Valembois et al., 1986).

Another hemolytic protein isolated from the coelomic fluid of *E. foetida* has been named eiseniapore (Lange et al., 1997). This protein is a thiol-activated hemolysin with a 38 kDa molecular weight and a strong lytic activity against erythrocytes. Experiments using lipid vesicles of various composition showed that eiseniapore requires the presence of sphingolipids.

Coelomic fluid of *E. foetida* has been shown to contain substances with cytolytic properties. A wide variety of cells including guinea pig cells, insect hemocytes and chicken fibroblasts has been found to be sensitive (Kauschke and Mohrig, 1987b). In *Lumbricus*, an active compound named lombricine has been purified (Nagasawa et al., 1991). Lombricine blocked the proliferation of spontaneous tumors in mice. Another substance with antitumor activities has been isolated from *E. foetida* and *L. rubelus* by Hrzenjak and co-authors (Hrzenjak et al., 1992). Despite the fact that the mechanism of the action of these antitumor substances is still unclear, this type of research will be more common in the near future.

The cytolytic activity of *E. foetida* coelomic fluid is not based on proteolysis and was found to be different from TNF-mediated lysis. Isolation and subsequent characterization of the cytolytic factor identified a 42-kDa cytolytic protein named coelomic cytolytic factor-1 (CCF-1) that exerts lytic activities on various mammalian cell lines (Bilej et al., 1995a). The authors hypothesize that CCF-1 might represent a primitive type of cytolytic vertebrate cytokine. Ultrastructural observations show multiple ruptures and defects in the erythrocyte and murine leukocyte membranes. Using other cell types, a disorganization of the macrophage microvilli, disorganization of cytoplasmic organelles and degranulation of mast cells were observed (Rossmann et al., 1997).

A biologically active glycolipoprotein complex with mitogenic properties on mammalian cells has been isolated from whole earthworm tissue extracts (Hrzenjak et al., 1993). Surprisingly, this complex reacted with anti-porcine insulin antibodies.

For detailed review of humoral defense mechanisms in annelids, see Bilej (1994b).

Adaptive humoral immunity

Laulan et al. were the first investigators analyzing the adaptively formed substances in earthworms. Using two synthetic haptens and two carrier proteins, they found that in response to immunization, *L. terrestris* synthesized specific molecules binding to the hapten-carrier complexes used for antimal stimulation. This response reached the maximum between days 5 to 8 after injection. Secondary immunization with the same complex resulted in faster response (Laulan et al., 1985). These results were later confirmed by Tučková and co-workers using different haptens (Tučková et al., 1988, 1991b). The newly formed protein was named antigen-binding protein (ABP) and was isolated and characterized. Its molecular weight is 56 kDa with two fractions (31 and 33 kDa, resp.) under reducing conditions.

Cellular immunity

Annelids respond to injury by a well-developed wound healing and regeneration. Using *Limnodrillus hoffmeisteri* as a model, Cornec and co-workers showed that stem cells supplied by each layer are involved in regeneration (Cornec et al., 1987). Intensive cell migration was observed during healing processes. For a comprehensive survey of regeneration and neurohormal control of these processes, see Bilej (1994a).

Phagocytosis

The first report of phagocytosis in annelids was done by Cameron (Cameron, 1932). He showed that all four types of tested inert material were readily phagocytized both *in vivo* and in *vitro* and that all types of coelomocytes with the exception of chloragogen cells are phagocytic. Using *L. terrestris* as a model, these results were later confirmed by Stein et al. (Stein et al., 1977). The maximum uptake was reached around 72–96 h. When synthetic polymeric microspheres were used *in vivo*, the phagocytosis was rapid with the maximal level only 3 h after injection. The main phagocytosing cells were the granulocytes (52% positive), followed by basophils (12%), acidophils (11%), and neutrophils (2%) (Bilej et al., 1990a).

Phagocytosis of bacteria was also described in the pioneering work of Cameron (Cameron, 1932). Using 21 different strains of bacteria, he demonstrated that most of the strains were phagocytosed rapidly, within 3 h. The most active cells were basophils. Subsequent studies confirmed these data (Dales and Kalac, 1992). Due to their lifestyle, the coelomic fluid of earthworms is not aseptic and is always contaminated with bacteria (Dales and Kalac, 1992). However, the rather high levels of phagocytes present in coelomic fluid together with humoral antibacterial substances can limit the numbers and growth of bacteria.

When yeast or sheep red blood cells were used as prey, a high rate of phagocytosis with a 60 min maximum was observed (Stein and Cooper, 1981; Laulan et al., 1988). This type of phagocytosis was significantly enhanced by opsonization with IgG or C3b fragment of complement. More detailed experiments revealed no role of IgM and C3d fragment. The reports about the opsonic effects of coelomic fluid are confusing. Some groups found no effects (Cooper, 1973), while others found higher level of internalization by neutrophilic coelomocytes (Stein and Cooper, 1981). One possible explanation might be that due to the low numbers of neutrophils, their level of phagocytosis might "disappear" among other, more frequent cell types. Using synthetic microspheres, Bilej and co-workers found that granulocytes are the prevailing phagocytic cell population and these particles opsonized in mouse serum are more rapidly phagocytosed (Bilej et al., 1990c). In contrast to some previous findings, opsonization in coelomic fluid also resulted in increased phagocytosis. Moreover, synthetic particles opsonized in antigen-stimulated worms were phagocytozed even faster. Subsequent experiments found two proteins (62 and 68 kDa) responsible for the opsonization of these particles (Sinkora et al., 1993).

When phagocytosis of synthetic particles was measured *in vitro* and *in vivo* in earthworms (*E. foetida or L. terrestris*) previously stimulated with antigen, the phagocytic activity was increased, with a maximum on day 4 after stimulation (Bilej et al., 1990c).

Beside mechanical ways of elimination of phagocytozed material, intracellular degradation plays an important role. The ability of earthworms to intracelullarly digest foreign proteins has been described already in the 1970s (Valembois et al., 1973) and has been further studied both *in vitro* and *in vivo* (Rejnek et al., 1993; Bilej et al., 1993). Despite high phagocytic activity, no oxidative burst has been detected (Bilej et al., 1991a), which might be caused by the use of synthetic inert material.

Larger particles are eliminated by encapsulation. The major role in encapsulation is played by a fibrous capsule called brown body (Valembois et al., 1992). Brown body contains two important pigments, lipofucin and melanin. Their function might be the inactivation of free radicals released to segregate unwanted particles (Valembois et al., 1994).

Cytotoxic reactions
Mammalian tumor cells K562 are killed by autogeneic but not allogeneic earthworm effector cells (Cooper et al., 1995; Suzuki and Cooper, 1995; Cossarizza et al., 1996). On the other hand, spontaneous *in vitro* cytotoxicity was observed in allogeneic cultures of *E. foetida* cells (Valembois et al., 1980). Earthworm nonphagocytic coelomocytes possess a high level of cytotoxicity against human NK-resistant and NK-sensitive targets (Cossa-

rizza et al., 1996b). Using human monoclonal antibodies, these cells were positive for CD11a, CD45RA, CD45RO,CDw49b, CD54 and Thy-1 antigens. Based on these interesting results, the authors postulate that primitive NK-like activity appeared early in evolution.

Transplantation reaction
Readers seeking detailed information about transplantation immunity in annelids should read the excellent and comprehensive reviews by Cooper and Roch (1994) and Cooper (1996).

Adaptive cellular defense reactions
Earthworms respond to an antigeneic challenge by active formation of an antigen-binding protein (ABP) (Tučková et al., 1988, 1991b). The ABP is involved not only in humoral defense reactions, but also plays a role in the cellular branch of annelid immunity. It is expressed on the membrane of coelomocytes, as was demonstrated by radioactive labeling (Bilej et al., 1990b, 1991b) and flow cytometry (Tučková et al., 1991a). The immunization not only increased the percentage of ABP-positive cells, but also the number of ABP on cell surface. Labeled antigens injected into the coelomic cavity are cleared within 5 days (Rejnek et al., 1993). However, almost 80% of antigen was cleaved into nonprecipitable fragment after 24 h. Comparing the fast destruction with the rather slow release of ABP, one can ask how this response is induced. Studies using radioactively labeled antigen revealed a homogenous localization during the first 24 h, after that the antigen was found only in chloragogenic tissue. Even if free coelomocytes did not show any signs of increased proliferation, the precursor cells in the mesenchymal lining did. These results were supported by incubation of pieces of chloragogenic tissue *in vitro* (Tučková et al., 1995). When the earthworms were stimulated *in vivo* with different proteins and the protein-binding was evaluated *in vitro* on both cellular and humoral level, the binding was found to be significantly higher when the same protein was used for stimulation. Moreover, the degree of specificity increased after the secondary *in vivo* challenge, but still was considerably lower that that of immunoglobulin binding specificity (Bilej et al., 1995b).

Coelomocytes are considered to be a terminal population, therefore their *in vitro* cultivation is almost impossible. A new approach to study the antigeneic response is the *in vitro* cultivation of gut wall tissue explants. These explants respond to antigen challenge by secretion of ABP into the culture medium (Bilej et al., 1994).

Analyses of the ABP synthesis revealed that in both the coelomic fluid and in coelomocytes a powerful proteolytic system is present. To answer the important question of how the rapid digestion of injected antigen is related to the ABP response, Tučková and Bilej tested the effects of proteolytic fragments on stimulation and subsequent ABP synthesis. In an elegant design they added the intact antigen as well as its proteolytic fragments to tissue explant cultures and measured the ABP response by indirect ELISA assays (Tučková and Bilej, 1994). They found that the response to small fragments was identical to that induced by the whole protein antigen. In addition, the response to the intact antigen was almost completely blocked by serine protease inhibitors, while the response to small fragments was only slightly decreased. These data represent the first direct proof of antigen processing in annelids.

Conclusions

The immune system of annelids may be clearly defined as the two-compound system developed from and housed in the coelom in which cooperate the cellular (coelomocytes) and humoral (coelomic fluid) part of immunity. Various coelomocyte types play a similar role like vertebrate lymphocytes, leucocytes and macrophages, whereas the agglutinating, opsonizing, and lytic factors secreted into coelomic fluid are vectors of humoral immunity. Both parts of annelide immunity are sufficiently effective in disposing of foreign material and pathogenic invaders as to be equivalent to the immune responses found in more advanced vertebrates. Moreover, the annelids, or more precisely the oligochaetes, are capable of an adaptive immune response by formation of humoral factors which are dissimilar with immunoglobulins and play a substitutive role of antibodies in these animals. Nevertheless, according to molecular genetics studies, some proteins belonging to the immunoglobulin superfamily could be present (Takahashi et al., 1995).

References

Barnes, R.D. (1987) *Invertebrate Zoology*. Saunders College Publishing, Philadelphia.

Barnes, R.D. (1989) Diversity of organisms: How much do we know? *Amer. Zool.* 29:1075–1084.

Bilej, M. (1994a) Cellular defense mechanisms. *In:* V. Větvička, P. Šíma, E.L. Cooper, M. Bilej, and P. Roch (eds): *Immunology of Annelids*, CRC Press, Boca Raton, pp 167–200.

Bilej, M. (1994b) Humoral defense mechanisms. *In:* V. Větvička, P. Šíma, E.L. Cooper, M. Bilej, and P. Roch (eds): *Immunology of Annelids*, CRC Press, Boca Raton, pp 245–261.

Bilej, M., Scheerlinck, J.-P., VandenDriessche, T., De Baetselier, P. and Větvička, V. (1990a) The flow cytometric analysis of *in vitro* phagocytic activity of earthworm coelomocytes (*Eisenia foetida:* Annelida). *Cell Biol. Int. Rep.* 14:831–837.

Bilej, M., Tučková, L., Rejnek, J. and Větvička, V. (1990b) *In vitro* antigen-binding properties of coelomocytes of *Eisenia foetida* (Annelida). *Immunol. Lett.* 26:183–188.

Bilej, M., Větvička, V., Tučková, L., Trebichavsky, I., Koukal, M. and Šíma P. (1990c) Phagocytosis of synthetic particles in earthworms. Effect of antigenic stimulation and opsonization. *Fol. Biol.* 36:273–280.

Bilej, M., De Baetselier, P., Trebichavsky, I. and Větvička, V. (1991a) Phagocytosis of synthetic particles in earthworms: absence of oxidative burst and possible role of lytic enzymes. *Fol. Biol.* 37:227–233.

Bilej, M., Rossmann, P., VandenDriessche, T., Scheerlinck, K.-P., De Baetselier, P., Tučková, L., Větvička, V. and Rejnek, J. (1991b) Detection of antigen in the coelomocytes of the earthworm, *Eisenia foetida* (Annelida). *Immunol. Lett.* 29:241–246.

Bilej, M., Tučková, L. and Rejnek, J. (1993) The fate of protein antigen in earthworms: study *in vitro. Immunol. Lett.* 35:1–6.
Bilej, M., Tučková, L. and Rossmann, P. (1994) A new approach to *in vitro* studies of antigenic response in earthworms. *Dev. Comp. Immunol.* 18:363–367.
Bilej, M., Tučková, L. and Romanovsky A. (1995b) Characterization of the limited specificity of antigen recognition in earthworms. *Fol. Microbiol.* 40:436–440.
Bilej, M., Brys, L., Beschin, A., Lucas, R., Vercauteren, E., Hanusova, R. and De Baetselier, P. (1995a) Identification of a cytolytic protein in the coelomic fluid of *Eisenia foetida* earthworms. *Immunol. Lett.* 45:123–128.
Brinkhurst, R.O. (1991) Ancestors (*Oligochaeta*), *Mitt. Hamb. Zool. Mus. Inst.* 88:97–110.
Cameron, G.R. (1932) Inflammation in earthworms. *J. Pathol. Bacteriol.* 35:933–972.
Clark, R.B. (1964) *Dynamics in Metazoan Evolution,* Claredon Press, Oxford.
Cooper, E.L. (1973) Evolution of cellular immunity. *In:* W. Braun and J. Ungar (eds): *Non-Specific Factors Influencing Host Resistance,* S. Karger, Basel, pp 11–23.
Cooper, E.L. (1996) Earthworm immunity. *In:* B. Rinkevich and W.E.G. Müller (eds): *Invertebrate Immunology,* Springer-Verlag, Berlin, pp 10–45.
Cooper, E.L. and Roch, P. (1994) Immunological profile of annelids: transplantation immunity. *In:* V. Větvička, P. Šíma, E.L. Cooper, M. Bilej, and P. Roch (eds): *Immunology of Annelids,* CRC Press, Boca Raton, pp 201–243.
Cooper, E.L. and Stein, E.A. (1981) Oligochaetes. *In:* N.A. Ratcliffe and A.F. Rowley (eds): *Invertebrate Blood Cells, Vol. 1,* Academic Press, London, pp 75–140.
Cooper, E.L., Lemmi, C.A.E. and Moore, T.C. (1974) Agglutinins and cellular immunity in earthworms. *Ann. NY Acad. Sci.* 234:34–49.
Cooper, E.L., Cossarizza, A., Suzuki, M.M., Salvioli, A., Capri, M., Quaglino, D. and Franceschi, C. (1995) Autogeneic but not allogeneic earthworm effector coelomocytes kill the mammalian tumor cell target K562. *Cell. Immunol.* 166:113–122.
Cornec, J.P., Cresp, J., Delye, P., Hoarau, F. and Reynaud, G. (1987) Tissue responses and organogenesis during regeneration in the oligochete *Limnodrilus hoffmeisteri* (Clap.). *Can. J. Zool.* 65:403–414.
Cossarizza, A., Cooper, E.L., Suzuki, M.M., Salvioli, S., Capri, M., Gri, G., Quaglino, D. and Franceschi, C. (1996) Earthworm leukocytes that are not phagocytic and cross-react with several human epitopes can kill human tumor cell lines. *Exp. Cell Res.* 224:174–182.
Dales, R.P. and Dixon, L.R.J. (1981) Polychaetes, *In:* N.A. Ratcliffe and A.F. Rowley (eds): *Invertebrate Blood Cells, Vol. 1,* Academic Press, London, pp 35–74.
Dales, R.P. and Kalac, Y. (1992) Phagocytic defence by the earthworm *Eisenia foetida* against certain pathogenic bacteria. *Comp. Biochem. Physiol. [A]* 101:487–490.
Darwin. C.R. (1881) The formation of vegetable mould through the action of worms with observations on their habits, John Murray and Co., London, pp 148.
Du Pasquier, L. and Duprat D. (1968) Aspects humoraux et cellulaires d'une immunite naturelle non specifiques chez l'Oligochete *Eisenia fetida (Oligochaeta). C. R. Acad. Sci. Paris* 266:538–541.
Friedman, M.M. and Weiss, L. (1982) The leucocytic organ of the megascolecid earthworm *Amynthas diffringens* (Annelida, Oligochaeta). *J. Morphol.* 174:251–268.
Graham, G.J. (1995) Tandem genes and clustered genes. *J. Theor. Biol.* 175:71–87.
Hill, R.L., Delaney, R., Fellows, R.E. Jr. and Lebovitz, H.E. (1966) The evolutionary origins of the immunoglobulins. *Proc. Natl. Acad. Sci. U.S.A.* 56: 1762–1769.
Hrzenjak, T., Hrzenjak, M., Kasuba, V., Efenberger-Marinculic, P. and Levanat, S. (1992) A new source of biologically active compounds-earthworm tissue (*Eisenia foetida, Lumbricus rubelus). Comp. Biochem. Physiol. [A]* 102:441–447.
Hrzenjak, M., Kobrehel, D., Levanat, S., Jurin, M. and Hrzenjak, T. (1993) Mitogenicity of the earthworm's (*Eisenia foetida*) insulin-like proteins. *Comp. Biochem. Physiol [B]* 104:723–729.
Kauschke, E. and Mohrig, W. (1987a) Comparative analysis of hemolytic and hemagglutinating activities in the coelomic fluid of *Eisenia foetida* and *Lumbricus terrestris, (Annelida, Lumbricidae). Dev. Comp. Immunol.* 11:331–341.
Kauschke, E. and Mohrig, W. (1987b) Cytotoxic activity in the coelomic fluid of the annelid *Eisenia foetida* Sav. *J. Comp. Physiol. [B]* 157:77–84.
Kauschke, E., Pagliara, P., Stabili, L. and Cooper, E.L. (1997) Characterization of proteolytic activity in coelomic fluid of *Lumbricus terrestris* L. (Annelida, Lumbricidae). *Comp. Biochem. Physiol. [B]* 116:235–242.

Lange, S., Nussler, F., Kauschke, E., Lutsch, G., Cooper, E.L. and Herrmann, A. (1997) Inter-
    action of earthworm hemolysin with lipid membranes requires sphingolipids. *J. Biol. Chem.*
    272:20884–20892.
Lassagues, M., Milochau, A., Doignon, F., Du Pasquier, L. and Valembois, P. (1997) Sequence
    and expression of an *Eisenia-fetida*-derived cDNA clone that encodes the 40-kDa fetidin
    antibacterial protein. *Eur. J. Biochem.* 246:756–762.
Laulan, A., Morel, A., Lestage, J., Dellage, M. and Chateaureynaud-Duprat, P. (1985) Evidence
    of synthesis by *Lumbricus terrestris* of specific substances in response to an immunization
    with a synthetic haptens. *Immunology* 56:751–758.
Laulan, A., Lestage, J., Bouc, A.M. and Chateaureynaud-Duprat, P. (1988) The phagocytic
    activity of *Lumbricus terrestris* coelomocytes is enhanced by the vertebrate opsonins: IgG
    and complement C3b fragment. *Dev. Comp. Immunol.* 12:269–278.
Mohrig, W. and Kauschke, E. (1984) Rosette formation of the coelomocytes of the earthworm
    Lumbricus terrestris L. with sheep erythrocytes. *Dev. Comp. Immunol.* 8:471–476.
Mohrig, W., Eue, I., Kauschke, E. and Hennicke, F. (1996) Crossreactivity of hemolytic and
    hemagglutinating proteins in the coelomic fluid of *Lumbricidae (Annelida)*. *Comp. Biochem.
    Physiol. [A]* 115:19–30.
Nagasawa, H., Sawaki, K., Fujii, Y., Kobayashi, M., Segawa, T., Suzuki, R. and Inatomi, H.
    (1991) Inhibition by lombricine from earthworm (*Lumbricus terrestris*) of the growth of
    spontaneous mammary tumours in SHN mice. *Anticanc. Res.* 11:1061–1064.
Ohno, S. (1970) *Evolution by Gene Duplication.* Springer, New York.
Peaucellier, G. (1983) Purification and characterization of proteases from the polychaete anne-
    lid *Sabellaria alveolata L. Eur. J. Biochem.* 136:435–445.
Rejnek, J., Tučková, L., Síma, P. and Bilej, M. (1993) The fate of protein antigen in earthworms:
    study *in vivo. Immunol. Lett.* 36:131–136.
Roch, P. (1979) Protein analysis of earthworm coelomic fluid: 1. Polymorphic system of the
    natural hemolysin of *Eisenia fetida andrei. Dev. Comp. Immunol.* 3:599–608.
Roch, P., Davant, N. and Lassegues, M. (1984) Isolation of agglutinins from lysins in earth-
    worm coelomic fluid by gel filtration followed by chromatofocusing. *J. Chromatogr.* 290:
    231–235.
Roch, P., Canicatti, C. and Valembois, P. (1989) Interactions between earthworm hemolysins
    and sheep red blood cell membranes. *Biochim. Biophys. Acta* 983:193–198.
Roch, P., Stabili, L. and Pagliara, P. (1991) Purification of three serine proteases from the coe-
    lomic cells of earthworms (*Eisenia fetida*). *Comp. Biochem. Physiol. [B]* 98:597–602.
Rossmann, P., Bilej, M., Tučková, L., Stary, V. and Kofronova, O. (1997) Lesion of leukocytes,
    erythrocytes, and mesothelial cells by the coelomic fluid of *Eisenia foetida* earthworm. *Fol.
    Microbiol.* 42:409–416.
Sawyer, R.T. and Fitzgerald, S.W. (1981) Hirudineans. *In:* N.A. Ratcliffe and A.F. Rowley (eds):
    Invertebrate Blood Cells, Vol. 1, Academic Press, London, pp 141–159.
Síma, P. (1994a) A survey of the evolution of fundamental body constructions in relation to
    immunological phenomena. *In:* V. Větvička, P. Síma, E.L. Cooper, M. Bilej, and P. Roch
    (eds): *Immunology of Annelids,* CRC Press, Boca Raton, pp 27–39.
Síma, P. (1994b) Phylogeny and classification of annelids. *In:* V. Větvička, P. Síma, E.L.
    Cooper, M. Bilej, and P. Roch (eds): *Immunology of Annelids,* CRC Press, Boca Raton,
    pp 13–25.
Síma, P. (1994c) Annelid coelomocytes and hemocytes: roles in cellular immune reactions.
    In: V. Větvička, P. Síma, E.L. Cooper, M. Bilej, and P. Roch (eds): *Immunology of Annelids,*
    CRC Press, Boca Raton, pp. 115–165.
Singer, S.J. and Doolitle, R.F. (1966) Antibody active sites and immunoglobulin molecules.
    Science 153:13–25.
Sinkora, M., Bilej, M., Tučková, L. and Romanovsky, A. (1993) Hemolytic function of opson-
    izing proteins of earthworm's coelomic fluid. *Cell Biol. Int.* 10:935–939.
Sinkora, M., Bilej, M., Drbal, K. and Tučková, L. (1995) Hemolytic function of opsonin-like
    molecules in coelomic fluid of earthworms. *Adv. Exp. Med. Biol.* 371:341–342.
Stein, E.A. and Cooper, E.L. (1981) The role of opsonins in phagocytosis by coelomocytes of
    the earthworm *Lumbricus terrestris. Dev. Comp. Immunol.* 5:415–425.
Stein, E.A and Cooper, E.L. (1983) Carbohydrate and glycoprotein inhibitors of naturally
    occurring and induced agglutinins in the earthworm *Lumbricus terrestris. Comp. Biochem.
    Physiol. [B]* 76:197–206.

Stein, E.A., Avtalion, R.R. and Cooper, E.L. (1977) The coelomocytes of the earthworm Lumbricus terrestris. *J. Morphol.* 153:467–476.

Stein, E.A., Younai, S. and Cooper, E.L. (1986) Bacterial agglutinins of the earthworm, *Lumbricus terrestris. Comp. Biochem. Physiol. [B]* 84:409–416.

Stein, E.A., Younai, S. and Cooper, E.L. (1990) Separation and partial purification of agglutinins from coelomic fluid of the earthworm, *Lumbricus terrestris. Comp. Biochem. Physiol. [B]* 97:701–705.

Stephenson, J. (1924) On the blood glands of earthworms of the genus *Pheretima. Proc. R. Soc. B.* 97:177–209.

Suzuki, M.M. and Cooper, E.L. (1995) Allogeneic killing by earthworm effector cells. *Nat. Immun.* 14:11–19.

Takahashi, T., Iwase, T., Kobayashi, K., Rejnek, J., Mestecky, J. and Moro, I. (1995): Phylogeny of the immunoglobulin joining (J) chain. *In:* J. Mestecky, M.W. Russel, S. Jackson, S.M. Michalek, H. Tlaskalova-Hogenova, and J. Sterzl (eds): *Advances in Mucosal Immunity, Part A,* Plenum Press, New York, pp 353–356.

Tučková, L. and Bilej, M. (1994) Antigen processing in earthworms. *Immunol. Lett.* 41: 273–277.

Tučková, L. and Bilej, M. (1996) Mechanisms of antigen processing in invertebrates: Are there receptors? *In:* E.L. Cooper (ed.): *Advances in Comparative and Environmental Physiology, Vol. 23, Invertebrate Immune Responses: Cells and Molecular Products,* Springer-Verlag, Berlin, pp 41–72.

Tučková, L., Rejnek, J. and Šíma, P. (1988) Response to parenteral stimulation in earthworms *L. terrestris* and *E. foetida. Dev. Comp. Immunol.* 12:287–296.

Tučková, L., Rejnek, J., Šíma, P. and Ondrejova, R. (1986) Lytic activities in coelomic fluid of *Eisenia foetida* and *Lumbricus terrestris. Dev. Comp. Immunol.* 10:181–189.

Tučková, L., Rejnek, J., Bilej, M., Hajkova, H. and Romanovsky, A. (1991a) Monoclonal antibodies to antigen binding protein of annelids (*Lumbricus terrestris*). *Comp. Biochem. Physiol. [B],* 100:19–23.

Tučková, L., Rejnek, J., Bilej, M. and Pospisil, R. (1991b) Characterization of antigen-binding protein in earthworms *Lumbricus terrestris* and *Eisenia foetida. Dev. Comp. Immunol.* 15: 263–268.

Tučková, L., Bilej, M. and Rejnek J. (1995) The fate of protein antigen in Annelids – *in vivo* and *in vitro studies. In:* J. Mestecky, M.W. Russel, S. Jackson, S.M. Michalek, H. Tlaskalova-Hogenova, and J. Sterzl (eds): *Advances in Mucosal Immunity, Part A,* Plenum Press, New York, pp 335–339.

Valembois, P., Roch, P. and Du Pasquier, L. (1973) Degradation *in vitro* de proteine entrangere par les macrophages du Lombricien *Eisenia fetida* Sav. *C. R. Acad. Sci. Paris Ser. III* 277:57–60.

Valembois, P., Roch, P. and Boiledieu, D. (1980) Natural and induced cytotoxicities in *Sipunculids* and *Annelids. In:* M.J. Manning (ed.): *Phylogeny of Immunological Memory,* Elsevier/North Holland, Amsterdam, pp 47–55.

Valembois, P., Roch, P., Lassagues, M. and Cassand, P. (1982) Antibacterial activity of the hemolytic system from the earthworm *Eisenia fetida andrei. J. Invertebr. Pathol.* 40:21–27.

Valembois, P., Roch, P. and Lassegues, M. (1986) Antibacterial molecules in annelids. *In:* M. Brehelin (ed.): *Immunity in Invertebrates,* Springer Verlag, Berlin, pp 74–93.

Valembois, P., Lassegues, M. and Roch, P. (1992) Formation of brown bodies in the coelomic cavity of the earthworm *Eisenia foetida andrei* and attendant changes in shape and adhesive capacity of constitutive cells. *Dev. Comp. Immunol.* 16:95–101.

Valembois, P., Seymour, J. and Lassegues, M. (1994) Evidence of lipofuscin and melanin in the brown body of the earthworm *Eisenia fetida andrei. Cell Tissue Res.* 277:183–188.

Vaillier, J., Cadoret, M.A., Roch, P. and Valembois, P. (1985) Protein analysis of earthworm coelomic fluid. III. Isolation and characterization of several bacteriostatic molecules from *Eisenia fetida andrei. Dev. Comp. Immunol.* 9:11–20.

Větvička, V., Šíma, P., Cooper, E.L., Bilej, M. and Roch, P. (1994) *Immunology of Annelids,* CRC Press, Boca Raton.

Voburka, Z., Maser, M., Větvička, V., Bilej, M., Baudys, M. and Fusek, M. (1992) New trypsin inhibitors are present in coelomic fluid of *Lumbricus terrestris. Biochem. Int.* 27:679–685.

Arthropoda

The arthropods are unequivocally the largest assemblage, comprising at least a million known species (Barnes, 1989; May, 1994), but on the basis of other conjectures, there may be as many as several millions of insect species living in the rainforest canopies alone (Erwin, 1983). The arthropods are metamerized eucoelomate animals like annelids, but the strong tendency to reduce the number of segments is a typical feature. The coelomic pouches in arthropods exist only in early embryological stages. Later, they fuse into a common cavity, the mixocoel (hemocoel) which forms an open blood circulation system. These novelties, together with the evolution of the chitinous exoskeleton unique to them, have brought vast changes not only in other morphological characters, but also the development of different non-morphological features reflecting both their immune strategy and way of life. Due to this, the arthropods represent the most succesful protostomate animals. Their enormous adaptibility allowed them to invade every habitat since their first appearance in the Precambrian and mainly in the Cambrian period.

There are many controversial hypotheses on the origin and kinship of arthropods to other main protostomial phyla, the annelids and molluscs, and even to vertebrates (for review see Anderson, 1973; Løvtrup, 1977; Brusca and Brusca, 1990). The traditional view based on the morphological similarities that annelids and arthropods could be regarded in close relationship to each other is at present supported by molecular studies and by Hox gene clusters comparative analyses (Kim et al., 1996; Valentine et al., 1996). It is beyond the purposes of this monography to discuss further hypotheses about if the arthropods form a monophyletic or polyphyletic assemblage (Manton, 1977). Accepting the conclusions of classical comparative anatomical views supported recently by molecular studies (Boore et al., 1995; Kim et al., 1996), we may consider the arthropods as a monophyletic group.

Most immunological comparative studies have been made only in several arthropod groupings; we regard the following subtaxa of arthropods as sufficient to survey the immune strategies of these animals: the subphylum *Chelicerata*, the subphylum *Crustacea*, and the class *Insecta* of the subphylum *Uniramia* (cf. Barnes, 1987). The information on the immune mechanisms of onychophorans and myriapods is still scarce, even if these uniramians comprise more than 10 000 known species. Their cellular and humoral immune vectors seem to be very similar to those of insects (for review see Ravindranath, 1981; Ratcliffe et al., 1982; Šíma and Větvička, 1990; Xylander and Nevermann, 1990; van der Walt and McClain, 1990; Attygale et al., 1993).

Despite the fact that the arthropods did not develop any immune strategy similar to the anticipatory adaptive immunity, they compensated this apparent incapability extremely successfully. They utilized all possibilities of their relatively simple body pattern determining their morphofunctional endowment to paraphrase inherited immune mechanisms and to invent evolutionary novelties which they transformed into new defensive weapons. Their successful ability of adaptive radiation allowing them to colonize every ecological niche must be due to the presence of a proper, well-developed immunity, effectively destroying almost all pathogenic attacks. Besides the "classical" mechanisms such as phagocytosis, encapsulation, clotting, cytotoxicity and graft rejection, they utilize their first advantage, the chitinous exoskeleton, which was changed into impenetrable barriers. Even if some species have developed immunity exerting a high degree of adaptive nature, we must realize that members of arthropod assemblage are taxonomically, evolutionary and therefore physiologically very dissimilar from each other.

References

Anderson, D.T. (1973) *Embryology and Phylogeny in Annelids and Arthropods*. Pergamon Press, Oxford.
Attygalle, A.B., Xu, S.C. and Meinwald, J. (1993) Defensive secretion of the millipede *Floridobolus penneri*. *J. Natural Products* 56:1700–1706.
Barnes, R.D. (1987) *Invertebrate Zoology, 5th Edition*. Saunders College Publ., Philadelphia.
Barnes, R.D. (1989) Diversity of organisms: How much do we know? *Amer. Zool.* 29:1075–1084.
Boore, J.L., Collins, T.M., Stanton, D., Daehler, L.L. and Brown, W.M. (1995) Deducing the pattern of arthropod phylogeny from mitochondrial DNA rearrangements. *Nature* 376:163–165.
Brusca, R.C. and Brusca, G.J. (1990) *Invertebrates*. Sinauer, Sunderland, MA.
Edwin, T.L. (1983) Beetles and other insects of tropical forest canopies at Manaus, Brazil, sampled by insecticidal fogging. *In:* S.L. Sutton, T.C. Whitmore and A.C. Chadwick (eds): *Tropical Rain Forest: Ecology and Management Vol. 2*, Blackwell Sci. Publ., Oxford, pp 59–75.
Kim, C.B., Moon, S.Y., Gelder, R.S. and Kim, W. (1996) Phylogenetic relationship of annelids, molluscs, and arthropods: evidence from molecules and morphology. *J. Mol. Evol.* 43:207–215.
Løvtrup, S. (1977) *The Phylogeny of Vertebrata*. J. Wiley and Sons, London.
Manton, S.M. (1977) *The Arthropoda: Habits, Functional Morphology, and Evolution*. Clarendon Press, Oxford.
May, R.M. (1994) Biological diversity: differences between land and sea, *Phil. Trans. R. Soc. London B* 343:105–111.
Ratcliffe, N.A., White, K.N., Rowley, A.F. and Walters, J.B. (1982) Cellular defense systems of the arthropoda. *In:* N. Cohen and M.M. Sigel (eds): *The Reticuloendothelial System, Vol. 3*, Plenum Press, New York, pp 167–255.
Ravindranath, M.H. (1981) Onychophorans and myriapods. *In:* N.A. Ratcliffe and A.F. Rowley (eds): *Invertebrate Blood Cells, Vol. 2*, Academic Press, London, pp 327–354.
Síma, P. and Větvička, V. (1990) *Evolution of Immune Reactions*. CRC Press, Boca Raton.
Valentine, J.W., Erwin, D.H. and Jablonski, D. (1996) Developmental evolution of metazoan bodyplans: the fossil evidence. *Dev. Biol.* 173:373–381.
van der Walt, R. and McClain, E. (1990) Phylogeny of arthropod immunity. An inducible humoral response in the Kalahari millipede, *Triaenostreptus triodus* (Attems). *Naturwissensch.* 77:189–190.
Zylander, W.E.R. and Neverman, L. (1990) Antibacterial activity in the hemolymph of myriapods *(Arthropoda)*. *J. Invertebr. Pathol.* 56:206–214.

Chelicerata

The *Chelicerata* comprise the classes *Merostomata*, the aquatic horseshoe crabs, the *Arachnida*, the largest assemblage including spiders, scorpions, pseudoscorpions, whip scorpions, amblypygids, solifuges, harvestmen, mites, and ticks, and the *Pycnogonia,* the sea spiders. The most distinguishing feature, common to all representatives of this subphylum, and differentiating them from remaining arthropod subtaxa, is the absence of antennae. The majority of immunological studies have been conducted on the horseshoe crabs, especially the *Limulus polyphemus*, and to a lesser extent on spiders and scorpions, humoral immunity of which remains a *terra incognita,* as does the immune capability of vast numbers of members of remaining arachnide subtaxa including the sea spiders.

The cells and structures engaged in the immune reactions

The only structures suspected to play a certain role in the horseshoe crab immunity seem to be hypodermal glands. They are situated below the external carapace. Stagner and Redmont (1975) have described the production of a glycoprotein substance functioning as a viscose mechanical barrier and having potent agglutinating and antibacterial activity by hypodermal gland cells in a response to bacterial endotoxin stimulation. This defensive scenario resembles that of an immune strategy found in the molluscs (see chapter on *Mollusca*, this volume).

The hemocytopoietic structures in scorpions have been reported to be the cell masses forming parts of lymph gland (or lymphoid organ) which is situated in the cephalothoracic and preabdominal regions or scattered throughout the body (Kollmann, 1910). In some species of scorpions, the hyaline and granular hemocytes have been described in the early 1990s as the main blood cell types (Kollmann, 1908). In the scorpion, *Palamnaeus swammerdami,* six basic types of blood cells including prohemocytes, plasmatocytes, granular hemocytes, cystocytes, spherule cells, and adipohemocytes have been indentified (Ravindranath, 1974). All these cells are believed to participate in the immune reactions, mainly in blood clotting and phagocytosis (Sherman, 1981, Ratcliffe et al., 1982).

In ticks, several types of hemocytes resembling those of insects have been identified up to now (Brington and Burgdorfer, 1971; Binnington and Obenchain, 1982; El Shoura, 1986, 1989, Kuhn and Haug, 1994). For the time being there is only one reference describing the participation of the tick hemocytes in the encapsulation reaction in *Dermacentror andersoni* (Eggenberger et al., 1990). In spiders, the hemocytopoietic tissue is localized in the heart wall (Franz, 1904; Seitz, 1972). The differentiation of hemocytes has been described in only a few species of arachnids (Deevey, 1941; Seitz,1972; for review see Sherman, 1981).

Humoral immunity

The most commonly studied species is *Limulus polyphemus* with the well described lectin called limulin (Roche and Monsigny, 1974; Novak and Barondes, 1975; Bishayee and Dorau, 1980). The first line of defense is, however, the product of the hypodermal gland, which provides a barrier against bacteria and has strong antibacterial effects (Stagner and Redmond, 1975).

Limulin is the best known lectin of all invertebrate lectins. It has a molecular weight of 330 kDa and its physicochemical properties (Roche and Monsigny, 1974, 1979) and amino acid composition (Kehoe et al., 1979) are known. For an excellent summary, see Ey and Jenkin (1982). Another family of lectins, the liphemin family, has been isolated from the hemolymph of both the American horseshoe crab *L. polyphemus* and the Asian horseshoe crab, *Tachypleus tridentatus* (Tsuboi et al., 1993a, b). Of interest is that all of these lectins require Ca^{2+} for agglutination of erythrocytes. A novel lectin with a MW of 1700 kDa consisting of 24 identical subunits with agglutinating activity against human, rabbit and horse erythrocytes has been isolated (Tsuboi et al., 1996). This lectin has no Ca^{2+} requirements.

Several lectins have been isolated from the Asian horseshoe crab, *T. tridentatus*. Tsuboi et al. isolated an agglutinin with a 533 kDa molecular weight (Tsuboi et al., 1993b). The lectin agglutinated horse erythrocytes and reacts specifically with sialic acids containing glycoprotein. Another lectin was isolated by Fischer and colleagues (1994). It is smaller, with a MW of 130 kDa and it consists of four units.

Several authors found that homologues of two plasma proteins of vertebrates, α_2-macroglobulin and C-reactive protein, participate in a hemolytic system of the *L. polyphemus* (Enghild et al., 1990, Armstrong, 1991). In *Limulus,* α_2-macroglobulin serves as a broad spectrum protease-binding molecule mediating the clearance of proteases from the plasma. C-reactive protein is an essential part of the hemolytical system. Purified C-reactive protein was hemolytic in the absense of α_2-macroglobulin (Armstrong et al., 1993). These observations strongly suggest that these proteins present in *L. polyphemus* plasma show evolutionary affinities with the vertebrate complement system. Pentraxins, important mediators of an immune response were isolated from the plasma of *L. polyphemus* (Armstrong et al., 1996). Two different pentraxins were found – *Limulus* C-reactive protein with unknown function, and limulin, mediating a cytolytic reaction resulting in foreign cell destruction (for review see Armstrong et al., 1996).

An anti-LPS factor isolated from the *T. tridentatus* amoebocytes was shown to have antibacterial activity (Morita et al., 1985). In addition, the tachyplesin peptides isolated from hemocytes of all four species of horseshoe crabs are highly effective against bacteria (Nakamura et al.,

1988; Muta et al., 1990). Another antimicrobial and antifungal factor was isolated from the plasma of the horseshoe crab, *Carcinoscorpius rotundicauda* (Yeo et al., 1993). It is different from the antibacterial lectins isolated from its serum and from the tachyplesins isolated from cells. For a detailed description of humoral immunity in the horseshoe crab, see Armstrong (1991).

Cellular immunity

In *Merostomata,* the cells responsible for defensive functions are the blood hemocytes and the hypodermal gland-derived cells, the only free cell type is the granular amoebocyte (granulocyte) (Loeb, 1902; Armstrong, 1979). On the other hand, some groups have found two kinds of free hemocytes: granylocytes and plasmatocytes in both the horseshoe crab *T. tridentatus* (Jakobsen and Suhr-Jessen, 1990) and *L. polyphemus* (Suhr-Jessen et al., 1989). Besides morphological differences, the plasmatocytes are not affected by LPS. The nomenclature of the basic cell types is still rather confusing. For a brief survey, see Šíma and Větvička (1990).

Different situation exists in the *Arachnida.* At least five or six morphologically distinctive cell types have been described (Ravindranath, 1974).

The coagulation in arthropods has been studied in greatest detail in the horseshoe crab, *L. polyphemus*, in which all proteins necessary for the clotting reaction are localized within the vesicles of the amoebocyte (Levin, 1967, 1985). Two distinct phases in this process can be distinguished: the agglutination of amoebocytes and the coagulation of the surrounding plasma, which is triggered by factors released from agglutinated cells and does not occur without LPS (Levin and Bang, 1964; Levin, 1967). The involvement of LPS was directly demonstrated by isolation of granules obtained from agglutinated cells, mixing them with LPS and subsequently triggering the clotting (Mürer et al., 1975).

Other defense reactions consist of encapsulation, clotting and phagocytosis (Armstrong, 1991), but the literature is scarce and the interpretation difficult (for a review, see Ratcliffe et al., 1982). Based on the scarce studies available, phagocytosis seems to be surprisingly low efficient and unimportant (Armstrong, 1991).

The clotting system is similar, if not identical, to that found in other arthropods. It is activated by either β-glucan or LPS through a cascade of reactions based on three kinds of serine proteases (Iwanaga et al., 1992). The end point of this cascade is marked by the formation of a gel which mobilizes the bacteria and thus prevents its spread (Shirodkar et al., 1960).

The phenoloxidase system was, on the other hand, supposed to be missing (Söderhäll et al., 1985; Armstrong, 1991). However, recent observations demonstrated the presence of this system in *L. polyphemus* (Nellaiappan and Sugumaran, 1996).

For a detailed description of cellular immunity of horseshoe crab, see Armstrong (1991).

References

Armstrong, P.B. (1979) Motility of the Limulus blood cells. *J. Cell Sci.* 37:169–170.

Armstrong, P.B. (1991) Cellular and humoral immunity in the horseshoe crab, *Limulus polyphemus. In:* A.P. Gupta (ed.): *Immunology of Insects and other Arthropods,* CRC Press, Boca Raton, pp 3–17.

Armstrong, P.B. and Levin, J. (1979) *In vitro* phagocytosis by *Limulus* blood cells. *J. Invertebr. Pathol.* 34:145–151.

Armstrong, P.B., Armstrong, M.T. and Quigley, J.P. (1993) Involvement of α_2-macroglobulin and C-reactive protein in a complement-like hemolytic system in the arthropod, *Limulus polyphemus. Mol. Immunol.* 30:929–934.

Armstrong, P.B., Melchior, R. and Quigley, J.P. (1996) Humoral immunity in long-lived arthropods. *J. Insect Physiol.* 42:53–64.

Binnington, K.C. and Obenchain, F.D. (1982) Structure and function of the circulatory, nervous, and neuroendocrine systems in ticks. *In:* F.D. Obenchain and R.L. Galun (eds): *Physiology of Ticks,* Pergamon Press, Oxford, pp 351–398.

Bishayee, S. and Dorai, D.T. (1980) Isolation and characterization of a sialic acid-binding lectin (carcinoscorpin) from Indian horseshoe crab *Carcinoscorpius rotunda cauda. Biochim. Biophys. Acta* 623:89–97.

Brinton, L.P. and Burgdorfer, W. (1971) Fine structure of normal hemocytes in *Dermacentror andersoni* Stiles *(Acari: Ixodidae). J. Parasitol.* 57:1110–1127.

Deevey, G.B. (1941) The blood cells of the Haitan tarantula and their relation to the moulting cycle. *J. Morphol.* 68:475–491.

Eggenberger, L.R., Lamoreaux, W.J. and Coons, L.B. (1990) Hemocytic encapsulation of implants in the tick *Dermatocentor variabilis. Exp. Appl. Acarol.* 9:279–287.

El Shoura, S.M. (1986) Fine structure of hemocytes and nephrocytes of *Argas (Persiciargas) arboreus (Ixodoidea: Argasidae). J. Morphol.* 189:17–24.

Enghild, J.J., Thogersin, I.B., Salvesen, G. Fey, G.H., Figler, N.L., Gonias, S.L. and Pizzo, S.V. (1990) α_2-Macroglobulin from *Limulus polyphemus* exhibits proteinase inhibitory activity and participates in a hemolytic system. *Biochemistry* 29:10070–10080.

Ey, P.L. and Jenkin, C.R. (1982) Molecular basis of self/non-self discrimination in the invertebrata. *In:* N. Cohen and M.M. Sigel (eds): *The Reticuloendothelial System, Vol. 3,* Plenum Press, New York, pp 321–391.

Fischer, E., Khang, N.Q., Letendre, G. and Brossmer, R. (1994) A lectin from the Asian horseshoe crab *Tachypleus tridentatus:* purification, specificity and interaction with tumour cells. *Glycoconjugate J.* 11:51–58.

Franz, V. (1904) Über die Structur des Herzens und die Entstehung von Blutzellen in Spinnen, *Zool. Anz.* 27:192–204.

Iwanaga, S., Miyata, T., Tokunaga, F. and Muta, T. (1992) Molecular mechanism of hemolymoh clotting system in *Limulus. Thrombosis Res.* 68:1–32.

Jakobsen, P.P. and Suhr-Jessen, P. (1990) The horseshoe crab *Tachypleus tridentatus* has two kinds of hemocytes: granulocytes and plasmocytes. *Biol. Bull.* 178:55–64.

Kehoe, J.M., Kaplan, R. and Steven, S.L.L. (1979) Functional implications of the covalent structure of limulin: An overview. *In:* E. Cohen (ed.): *Biomedical Applications of the Horseshoe Crab (Limulidae),* Alan R. Liss, New York, pp 617–630.

Kollmann, M. (1908) Recherches sur les leucocytes et le tissue lymphoide des Invertébrés. *Ann. Sci. Nat. Zool.* 8:1–240.

Kollmann, M. (1910) Notes sur les functions de la glande lymphatique des scorpionides. *Bull. Soc. Zool. France* 35:25–30.

Kuhn, K.H. and Haug, T. (1994) Ultrastructural, cytochemical, and immunocytochemical characterization of haemocytes of the hard tick *Ixodes ricinus, (Acari; Chelicerata). Cell Tissue Res.* 227: 93–504.

Levin, J. (1967) Blood coagulation and endotoxin in invertebrates. *Fed. Proc.* 26: 1707–1712.

Levin, J. (1985) The role of amebocytes in the blood coagulation mechanism of the horseshoe crab, *Limulus polyphemus. In:* W.D. Cohen (ed.): *Blood Cells of Marine Invertebrates, Vol. 6,* Alan R. Liss, New York, pp 145–163.

Levin, J. and Bang, F.B. (1964) The role of endotoxin in the extracellular coagulation of Limulus blood. *Bull Johns Hopkins Hosp.* 115:265–274.

Loeb, L. (1902) On the blood lymph cells and inflammatory processes of Limulus. *J. Med. Res.* 7:145–165.

Morita, T., Ohtsubo, S., Nakamura, T., Iwanaga, S., Ohashi, K. and Niwa, M. (1985) Isolation and biological activities of *Limulus* anticoagulant (anti-LPS) which interacts with lipopolysaccharide (LPS). *J. Biochem.* 97:1611–1620.

Mürer, E.H., Levin, J. and Holme, R. (1975) Isolation and studies of the granules of the amoebocytes of *Limulus polyphemus,* the horseshoe crab. *J. Cell Physiol.* 86:533–542.

Muta, T., Fujimoto, T., Nakajima, H. and Iwanaga, S. (1990) Tachyplesins isolated from haemocytes of Southeast Asian horseshoe crab (*Carcinoscorpius rotundicauda* and *Tachypleus gigas*): identification of a new tachyplesin, tachyplesin III, and a processing immediate of its precursors. *J. Biochem.* 108:261–266.

Nakamura, T., Furunaka, H., Miyata, T., Tokunaga, F., Muta, T. and Iwanaga, S. (1988) Tachyplesin, a class of antimicrobial peptides from the haemocytes of the horseshoe crab, *Limulus polyphemus. J. Biol. Chem.* 263:16709–16713.

Nellaiappan, K. and Sugumaran, M. (1996) On the presence of prophenoloxidase in the hemolymph of the horseshoe crab, *Limulus. Comp. Biochem. Physiol. [B]* 113:163–168.

Nowak, T.P. and Barondes, S.H. (1975) Agglutinin from *Limulus polyphemus. Biochim. Biophys. Acta* 393:115–123.

Ratcliffe, N.A., White, K.H., Rowley, A.F. and Walters, J.B. (1982) Cellular defense systems of the arthropoda. *In:* N. Cohen and M.M. Sigel (eds): *The Reticuloendothelial System, Vol. 3,* Plenum Press, New York, pp 167–215.

Ravindranath, M.H. (1974) The hemocytes of a scorpion *Palamnaeus swammerdami. J. Morphol.* 144:1–10.

Roche, A.C. and Monsigny, M. (1974) Purification and properties of limulin: A lectin (agglutinin) from hemolymph of *Limulus polyhemus. Biochim. Biophys. Acta* 371:242–254.

Roche, A.C. and Monsigny, M. (1979) Limulin (*Limulus polyphemus* lectin): Isolation, physicochemical properties, sugar specificity and mitogenic activity. *In:* E. Cohen (ed.): *Biomedical Applications of the Horseshoe Crab (Limulidae),* Alan R. Liss, New York, pp 603–616.

Seitz, K.A. (1972) Zur Histologie und Feinstruktur des Herzens und der Hämazyten von *Cupiennius salei* Keys (*Araneae, Ctenidae*). II. Zur Funktionsmorphologie der Phagocyten. *Zool. Jahrb. Anat.* 89:385–397.

Sherman, R.G. (1981) Chelicerates. *In:* N.A. Ratcliffe and A.F. Rowley (eds): *Invertebrate Blood Cells, Vol. 2,* Academic Press, London, pp 355–384.

Shirodkar, M.V., Warwick, A. and Bang, F.B. (1960) The *in vitro* reaction of *Limulus* amoebocytes to bacteria. *Biol. Bull.* 118:324–337.

Síma, P. and Větvička, V. (1990) *Evolution of Immune Reactions.* CRC Press, Boca Raton.

Söderhäll, K., Levin, J. and Armstrong, P.B. (1985) The effect of β-1,3 glucans on blood coagulation and amebocyte release in the horseshoe crab *Limulus polyphemus. Biol. Bull.* 169:661–674.

Stagner, J.I. and Redmond, J.R. (1975) The immunological mechanisms of the horseshoe crab, *Limulus polyphemus. Mar. Fish. Rev.* 37:11–19.

Suhr-Jessen, P., Baek, L. and Jakobsen, P.P. (1989) Microscopical, biochemical and immunological studies of the immune defense system of the horseshoe crab, *Limulus polyphemus. Biol. Bull.* 176:290–300.

Tsuboi, I., Matsukawa, M., Sato, N. and Kimura, S. (1993a) Isolation and characterization of a sialic acid-specific binding lectin from the hemolymph of Asian horseshoe crab, *Tachypleus tridentatus. Biochem. Biophys. Acta* 1156:255–262.

Tsuboi, I., Yanagi, K., Wada, K., Kimura, S. and Ohkuma, S. (1993b) Isolation and characterization of a novel sialic acid-specific lectin from hemolymph of *Limulus polyphemus. Comp. Biochem. Physiol. [B]* 104:19–26.

Tsuboi, I., Yanagi, K., Matsukawa, M., Kubota, H. and Yamakawa, T. (1996) Isolation of a novel lectin from the hemolymph of horseshoe crabs *Limulus polyphemus* and its hemagglutinating properties. *Comp. Biochem. Physiol. [B]* 113:137–142.

Yeo, D.S.A., Ding, J.L. and Ho, B. (1993) An antimicrobial factor from the plasma of the horseshoe crab, *Carcinoscorpius rotindicauda. Microbios* 73:45–58.

Crustacea

The crustaceans form the only arthropodan subphylum whose members are primarily aquatic. Typical crustacean features are the presence of two pairs of antennae and calcified cuticle. They are classified into a row of subtaxa comprising commonly known crabs, shrimps, water fleas, barnacles, lobsters, crayfish, and wood lice. Unlike the situation with the previous chelicerates, much more attention has been devoted to crustacean species by comparative immunologists.

Hemopoietic structures

Hemopoietic tissue in a form of distinct organs has been described in many crustaceans. Hemopoiesis takes place in a series of nodules surrounded by a sheath of connective tissue localized in various species most often in the vicinity of the opthalmic artery, at the base of rostrum, or spread within the dorsal and lateral walls of the foregut, and in epigastric tissue (Cuénot, 1897; Kollmann, 1908; Martin et al., 1993). In the nodules, the hemoblasts with large nuclei and mitotic events are considered to be stem cells and can be seen together with small hyaline cells which leave the organ and migrate throughout the body, giving rise to the mature granulocytes (Ghiretti-Magaldi et al., 1977). It was documented that the hemopoietic activity is under hormonal influence, or the numbers of circulating hemocytes depend on stress, starvation and ambient temperature (Drach, 1939; Bauchau and Plaquet, 1973; Charmantier-Daures, 1973; Hamann, 1975; Ravindranath, 1977b). No phagocytic activity or production of humoral immune substances have been observed in the hemopoietic organs in crustaceans (Ratcliffe et al., 1982). On the other hand, an active function in the clearance of protein antigenic substances and possibly small particles like viruses has been described in branchial podocytes (nephrocytes). These specialized cells are located in the gills of some decapod crustaceans, podocytes of antennal excretory gland, in fixed phagocytes within the hepatopancreatic hemal spaces, in the ventral side of the stomach (Cuenot's "phagocytic organ" and "sub-stomachal organ"), and heart (Cuénot, 1905; Fontaine and Laitner, 1974; Smith and Ratcliffe, 1980a; Johnson, 1987; Sagrista and Durfort, 1990).

Humoral immunity

The reports of bactericidins in crustaceans remain controversial. Some authors found the bactericidins in hemolymph (Paterson et al., 1976; Smith and Ratcliffe, 1978; Martin et al., 1993), other groups showed the presence of heat-labile bactericidins in the hemolymph of the American lobster, *Homarus americanus* (Stewart and Zwicker, 1974).

The prophenoloxidase (proPO) system is an important part of the immunological recognition mechanisms not only in crustaceans, but in all arthropods. Its activation is involved in elicitation of phagocytosis (Smith and Söderhäll, 1983), nodule and capsule formation (Kobayashi et al., 1990), melanin synthesis (Lackie, 1988) and hemocyte locomotion (Takle and Lackie, 1986). The proPO system is triggered by β-glucan and LPS and is blocked by serine protease inhibitors, such as leupeptin and PMSF (Lanz et al., 1993a). This system thus mediates several activities similar to complement system, e.g., cell lysis, opsonization and substance release. Therefore, some speculation that the proPO system of invertebrates fulfills functions similar to that of complement in the vertebrates (Söderhäll, 1982; Söderhäll and Smith, 1986). The fact that both are enzymatic cascades involving proteolytic cleavage, serine proteases, generation of opsonins and have Ca^{2+} requirements seems to further support this hypothesis.

Two independent factors are involved in recognition of microorganisms. One is LPS-binding protein (Vargas-Albores et al., 1993b) and the second one is β-glucan-binding protein (Duvic and Söderhäll, 1990). These proteins enhance the activity of the prophenoloxidase-activating enzyme, which then induce activation of phenoloxidase (Aspan and Söderhäll, 1991; for review see Söderhäll and Cerenius, 1992; Cerenius and Söderhäll, 1995; Söderhäll et al., 1994a, b). In shrimp glucans activate both hemocytes and the prophenoloxidase system (Vargas-Albores et al., 1993a). In brown shrimp, *Panaeus californiensis,* a β-glucan-binding protein has been identified and purified (Vargas-Albores et al., 1996). It is a monomeric protein with a molecular weight of 100 kDa with a strong amino acid sequence similarity to the β-glucan-binding protein of crayfish. In the shore crab *Carcinus maenas*, a β-glucan-binding protein has a molecular weight of 110 kDa. It has opsonic activity resulting in higher levels of phagocytosis (Thornqvist et al., 1994).

Natural hemolytic activity was detected in brown shrimp *P. californiensis* hemolymph. It was unrelated to phenoloxidase or agglutinating activity. The hemolytic factor was thermolabile and had a MW of 23.5 kDa (Guzman et al., 1993).

Two proteins which bear a structural resemblance to complement proteins have been found in crustaceans (for review see Cerenius and Söderhäll, 1995). One is α_2-macroglobulin (Spycher et al., 1987, Hall et al., 1989), which was found to contain a thioester region with significant homology to vertebrate α_2-macroglobulin. Complement factors C3, C4 and C5 contain similar thioester region. It has been suggested that α_2-macroglobulin could constitute an evolutionary forerunner to the complement proteins (Sottrup-Jensen et al., 1985).

Another example is the presence of a complement protein repeat in horseshoe crab factor C. There are five domains, each comprising about 60 amino acid residues. Each of these domains is considerably similar to the short consensus repeats present in several complement factors (Muta et al., 1991).

Lectins, on the other hand, have been extensively studied in this assemblage since the pioneering work of Cantacuzene (1912). Despite the fact that the physiological function remains unclear, the involvement in recognition and defense is beyond doubt (Chattopadhyay and Chatterjee, 1993; for review see Yeaton, 1981; Ratner and Vinson, 1983; Vasta and Marchalonis, 1984; Šíma and Větvička, 1990; Söderhäll and Cerenius, 1992). A limited number of studies focused on lectins in crayfish. In the freshwater crayfish *Pacifastatus leniusculus* lectin was isolated and characterized with an affinity for lipopolysaccharide (Kopacek et al., 1993).

To summarize these findings, a number of conclusions can be drawn. Lectins have been found in almost all species tested. Quite often, there is a complex array of molecules with similar specificities. The specificity of N-acetylneuraminic acid is a widespread characteristic (Vasta and Cohen, 1982).

Crustacean hemolymph is able to coagulate and is commonly used both in prevention of contamination and loss of hemolymph after puncture (Levin, 1967). This is particularly important as crustaceans have an open circulatory system. Another defensive approach is clotting of hemocytes as an initial step of coagulation. Hyaline cells that burst to initiate clot formation were identified in *Panulirus interruptus* and *Homarus americanus* (Hose et al., 1990; Martin et al., 1991). Further studies showed that the clot-initiating burst of hyaline and semigranular cells is specifically induced by factors released from granular cells (Aono et al., 1994a). When hemocytes isolated from *P. japonicus* were mixed with serum of the same species, rapid cytolysis occurred in hyaline and semigranular cells. The same results were obtained in two other species, *H. americanus* and *P. japonicus* (Aono et al., 1994b). Dialyzed plasma had the same effects. The cytotoxic activity of plasma was weakened by heat treatment and completely inactivated by proteases.

Cellular immunity

Extensive literature exists on the morphology of crustacean hemocytes (for a review see Ratcliffe et al., 1982; Söderhäll and Cerenius, 1992). The basic categories of crustacean hemolymph cells are listed in Table 2. Generally, between two and four basic types are found, but severe confusion still remains. The presence of various cells with intermediate properties makes the whole situation even more difficult. A firmly established classification does not exist. The most important cell types seem to be hyaline, semigranular and granular cells (Bauchau, 1986). These cells were originally thought to be specialized for carrying out specific physiological tasks, such as phagocytosis (hyaline cells), nodule formation and coagulation (semigranular cells). More recent studies, however, suggest some degree of cell cooperation in all these functions through the prophenoloxidase activating

Table 2. The basic categories of crustacean hemocytes

Category	Characteristics
Granulocytes	Oval shape Kidney-shaped nuclei Abundant granules Phagocytosis Encapsulation Coagulation
Semigranular cells	Oval or spindle-shaped Oval nucleus Active phagocytosis Wound repair Coagulation Encapsulation
Hyaline cells	Oval or round-shaped Central large nucleus Pseudopodia Non or few granules Phagocytosis Encapsulation Coagulation
Miscellaneous cells	

system (Söderhäll and Smith, 1986a, b; Aspan et al., 1990). Crustacean blood cells are sometimes considered as circulating hepatocytes for the storage and release of carbohydrates. However, more recent studies showed that hemocyte differentiation could be accompanied by a loss of glyco-genosynthesis capability (Loret, 1993). (Readers seeking more details on the morphology of hemocytes should read the following reviews: Gupta, 1979; Bauchau, 1981; Šíma and Větvička, 1992; Lanz et al., 1993b)

Similarly to all arthropods, nodule formation and hemocyte aggregation can be seen in crustaceans (Smith and Racliffe, 1980a). Both processes are involved in bacterial clearance, but the exact explanation of the mechanisms of bacterial killing is still missing (Söderhäll and Cerenius, 1992).

Phagocytosis is the major mechanism of cellular defense. Besides free hemocytes, two kinds of fixed cells, the so-called phagocytic reserve cells and the fixed phagocytes, are involved. The morphology and functions of these cells were described in an excellent and extensive review by Johnson (1987). The most active phagocytosing cells among free hemocytes are hyaline cells. The whole process of bacterial elimination is extremely fast, more than 97% of all bacteria were eliminated within 6 h (Smith and Ratcliffe, 1980b). More detailed studies showed that the association of bacteria with the cell membrane, followed by sequestration within the cellular aggregates, constitutes the most significant mechanism for the bacteria clearance (Smith and Ratcliffe, 1980b; Ratcliffe et al., 1982). Based on these observations, it was hypothesized that phagocytosis is the first line of

defense against smaller amounts of invading bacteria, whereas cell aggregation is the secondary line used in case of overwhelming numbers of in-
truders (Ratcliffe et al., 1982). For more information about crustaceans'
phagocytosis, see Šíma and Větvička (1990).

An interesting observation of the opsonin specificity and recognition
was done by Sloan et al. (1975) using the crayfish *Procambarus clarkii* as
a model. These authors tested the clearance of various ^{125}I-labeled proteins
and found some degree of self/non-self recognition. When a mixture of hot
and cold proteins was injected, the excess of other proteins was used to eva-
luate the recognition system. Based on these results, authors concluded that
several independent natural recognition molecules exist in the hemolymph
of this creature.

Martin et al. studied mechanisms of clearance of bacteria in the penaeid
shrimp, *Sicyonia ingentis*, and compared it to those of other crustaceans
(Martin et al., 1993). These authors found that while many of the clearance
mechanisms are common to the crustaceans, several differences exist. Like
in other decapods, all types of Gram-positive bacteria were eliminated, and
this removal is not associated with humoral bactericidins. These results are
in agreement with those of Smith and Ratcliffe (1978) and Paterson et al.
(1976). Bacterial clearance was associated with a rapid decline (up to 80%
decline) in the number of circulating hemocytes (Tyson and Jenkins, 1973).
Surprisingly, when a second bacterial challenge is given to a hemocyte-
deficient animal, the bacteria are cleared as effectively as before (Martin et
al., 1993). It takes 72 to 96 h to restore the numbers of hemocytes. Mitotic
indices of the hematopoietic tissues are elevated soon after the bacterial
injection. Hemolymph cells were originally assumed to lack the ability to
divide, as the hematopoiesis was supposedly limited to the hematopoietic
tissue (Martin and Hose, 1991; Hose et al., 1992), but the latest experiments using more sensitive flow cytometry technique found a small but
consistent percentage of circulating hemocytes (Sequeira et al., 1996).

Hyaline cells of the shore crab, *Carcinus maenas*, produce superoxide
ions as products of stimulation. Concanavalin A, phorbol myristate acetate
and LPS are among substances able to elicit a superoxide production (Bell
and Smith, 1993). These findings represent the first demonstration of a
respiratory burst for crustaceans.

Very few studies have been devoted to the cellular cytotoxic reaction.
Two studies demonstrated the cytotoxic reaction using tumor cell lines as
targets (Tyson and Jenkins, 1973; Söderhäll et al., 1985), but the mechanism responsible for this activity or the physiological significance remain
to be solved.

References

Aono, H., Diaz, G.G. and Mori, K. (1994a) Granular cells recognize non-self signals and trigger the clotting reaction of hemocytes *in vitro* in the spiny lobster, *Panulirus japonicus*. *Comp. Biochem. Physiol. [A]* 107:37–42.

Aono, H., Diaz, G.G. and Mori, K. (1994b) Cytolysis of hemocytes induced by serum and plasma in three crustacenas, *Panulirus japonicus, Panaeus japonicus* and *Homarus americanus*. *Dev. Comp. Immunol.* 18:265–275.

Aspan, A. and Söderhäll, K. (1991) Purification of peophenoloxidase from crayfish blood cells, and its activation by an endogenous serine proteinase. *Insect Biochem.* 21:363–373.

Aspan, A., Hall, M. and Söderhäll, K. (1990) The effect of endogenous proteinase inhibitors on the prophenoloxidase activating enzyme, a serine proteinase from crayfish haemocytes. *Insect Biochem.* 20:485–492.

Bauchau, A.G. (1981) Crustaceans. *In:* N.A. Ratcliffe and A.F. Rowley (eds): *Invertebrate Blood, Vol. 2*, Academic Press, London, pp 385–420.

Bauchau, A.G. (1986) Donnees recentes sur les hemocytes des crustaces. *Cah. Biol. Mar.* 28: 279–287.

Bauchau, A.G. and Plaquet, J.C. (1973) Variations du nombre des hémocytes chez les crustacés brachyoures. *Crustaceana* 24:215–223.

Bell, K.L. and Smith, V.J. (1993) *In vitro* superoxide production by hyaline cells of the shore crab *Carcinus maenas* (L.). *Dev. Comp. Immunol.* 17:211–219.

Cantacuzene, J. (1912) Sur certains anticorps naturels observes chez *Eupagurus prideauxii*. *Compt. Rendus Soc. Biol.* 73: 663–683.

Charmantier-Daures, M. (1973) Activité de l'organe léucopoiétique de *Pachygrapsus marmoratus* au counts du cycle d'internue: Influence possible chez hormones pedonaulaires. *Bull. Soc. Zool. France* 98:221–231.

Cerenius, L. and Söderhäll, K. (1995) Crustacean immunity and complement; a premature comparison? *Amer. Zool.* 35:60–67.

Chattopadhyay, T. and Chatterjee, B.P. (1993) A low-molecular weight lectin from the edible crab *Scylla serrata* hemolymph: purification and partial characterization. *Biochem. Arch.* 9: 65–72.

Cuénot, L. (1897) Les globules sanquins et les organes lymohoides des invertébrés (Revue critique et nouvelles recherches). *Arh. Anat. Microscop. Exp.* 1:153–192.

Cuénot, L. (1905) L'organe phagocytaire des Crustacés Décapodes. *Arch. Zool. Exp. Gen. Ser.* 4:1–16.

Drach, P. (1939) Mue at cycle d'intermue chez les crustacés décapodes. *Ann. Inst. Oceanogr.* 19:103–391.

Duvic, B. and Söderhäll, K. (1990) Purification and characterization of a β-1,3-glucan binding protein from the plasma of the crayfish *Pacifastacus lenieusculus*. *J. Biol. Chem.* 265: 9332–9337.

Fontaine, C.T. and Lightner, D.V. (1974) Observations on the phagocytosis and elimination of carmine particles injected into the abdominal musculature of the white shrimp, *Penaeus setiferus*. *J. Invertebr. Pathol.* 24:141–148.

Ghiretti-Magaldi, A., Milanesi, C. and Tognon, G. (1977) Haemopoiesis in *Crustacea, Decapoda*: Origin and production of haemocytes and cyanocytes of *Carcinus maenas*. *Cell Differ.* 6:167–186.

Gupta, A.P. (1979) Arthropod hemocytes and phylogeny. *In:* A.P. Gupta (ed.): *Arthropod Phylogeny*, Van Nostrand-Reinhold, New York, pp 669–679.

Guzman, M.-A., Ochoa, J.L. and Vargas-Albores, F. (1993) Haemolytic activity in the brown shrimp (*Penaeus californiensis* Holmes) haemolymph. *Comp. Biochem. Physiol. [A]* 106: 271–275.

Hall, M., Soderhäll, K. and Sottrup-Jensen, L. (1989) Amino acid sequence around the thioester of a α_2-macroglobulin from plasma of the crayfish *Pacifastacus leniusculus*. *FEBS Lett.* 254: 11–114.

Hamann, A. (1975) Stress induced changes in cell-titre in crayfish haemolymph. *Z. Naturforsch.* 30c:850.

Hose, J.E., Martin, G.G. and Gerard, A.S. (1990) A decapod hemocyte classification scheme integrating morphology, cytochemistry, and function. *Biol. Bull.* 178:33–45.

Hose, J.E., Martin, G.G., Tiu, S. and McKrell, N. (1992) Patterns of hemocyte production and release throughout the molt cycle in the penaeid shrimp *Sicyonia ingentis*. *Biol. Bull.* 183: 185–189.

Johnson, P.T. (1987) A review of fixed phagocytic and pinocytic cells of decapod crustaceans. with remarks on hemocytes. *Dev. Comp. Immunol.* 11:679–704.

Kobayashi, M., Johansson, M.W. and Söderhäll, K. (1990) The 76 kD cell-adhesion factor from crayfish hemocyte promotes encapsulation *in vitro. Cell Tissue Res.* 260:13–18.

Kollmann, M. (1908) Recherches sur les leucocytes et le tissue lymphoide des Invertébrés. *Ann. Sci. Nat. Zool.* 8:1–240.

Kopacek, P., Grubhoffer, L. and Söderhäll, K. (1993) Isolation and characterization of a hemagglutinin with affinity for lipopolysaccharide from plasma of the crayfish *Pacifastacus leniusculus. Dev. Comp. Immunol.* 17:407–418.

Lackie, A.M. (1988) Immune mechanism in insects. *Parasitol. Today* 4:98–105.

Lanz, H., Hernandez, S., Garrido-Guerrero, E., Tsutsumi, V. and Arechiga, H. (1993a) Phenoloxidase system activation in the crayfish *Procambarus clarki. Dev. Comp. Immunol.* 17: 399–406.

Lanz, H., Tsutsumi, V. and Arechiga, H. (1993b) Morphological and biochemical characterization of *Procambarus clarki* blood cells. *Dev. Comp. Immunol.* 17:389–397.

Levin, J. (1967) Blood coagulation and endotoxin in invertebrates. *Fed. Proc.* 26:1707–1712.

Loret, S.M. (1993) Hemocyte differentiation in the shore crab (*Carcinus maenas)* could be accompanied by a loss of glycogenosynthesis capability. *J. Exp. Zool.* 2267:548–555.

Martin, G.G. and Hose, J.E. (1991) Vascular elements of blood (hemolymph). *In:* F.W. Harrison and A.G. Humes (eds): *Microscopic Anatomy of Invertebrates,* Wiley-Liss, New York, pp 117–146.

Martin, G.G., Hose, J.E., Omori, S., Chong, C., Hoodboy, T. and McKrell, N. (1991) Localization and roles of coagulogen and transglutaminase in hemolymph coagulation in decapod crustaceans. *Comp. Biochem. Physiol. [B]* 100:517–522.

Martin, G.G., Poole, D., Poole, C., Hose, J.E., Arias, M., Reynolds, L., McKrell, N. and Whang, A. (1993) Clearance of bacteria injected into the hemolymph of the penaeid shrimp, *Sicyonia ingentis. J. Invertebr. Pathol.* 62:308–315.

Muta, T., Miyata, T., Misumi, Y., Tokunaga, F., Nakamura, T., Toh, Y., Ikehara, Y. and Iwanaga, S. (1991) An endotoxin sensitive serine protease zymogen with a mosaic structure of complement-like, epidermal growth factor-like and lectin like domains. *J. Biol. Chem.* 266: 6554–6561.

Paterson, W.D., Stewart, J.E. and Zwicker, B.M. (1976) Phagocytosis as a cellular immune mechanism in the American lobster, *Homarus americanus. J. Invertebr. Pathol.* 27: 95–104.

Ratcliffe, N.A., White, K.H., Rowley, A.F. and Walters, J.B. (1982) Cellular defense systems of the arthropoda. *In:* N. Cohen and M.M. Sigel (eds): *The Reticuloendothelial System, Vol. 3,* Plenum Press, New York, pp 167–255.

Ratner, S. and Vinson, R.B. (1983) Phagocytosis and encapsulation: Cellular immune response in *Arthropoda. Am. Zool.* 23:185–194.

Ravindranath, M.H. (1977) The circulating haemocyte population of the mole crab *Emerita (= Hippa) asiatica* (Milne-Edwards). *Biol. Bull.* 152:415–423.

Sagrista, E. and Durfort, M. (1990) Ultrastructural study of hemocytes and phagocytes associated with hemolymphatic vessels in the hepatopancreas of *Palaemonetes zariquieyi* (Crustacea, Decapoda). *J. Morphol.* 206:173–180.

Sequeira, T., Tavares, D. and Arala-Chaves, M. (1996) Evidence for circulating hemocyte proliferation in the shrimp *Penaeus japonicus. Dev. Comp. Immunol.* 20:97–104.

Šíma, P. and Větvička, V. (1990) *Evolution of Immune Reactions.* CRC Press, Boca Raton.

Šíma, P. and Větvička, V. (1992) Evolution of immune accessory functions. *In:* L. Fornůsek and V. Větvička (eds): *Immune System Accessory Cells,* CRC Press, Boca Raton, pp 1–55.

Sloan, B., Yocum, C. and Clem, L.W. (1975) Recognition of self from non-self in crustaceans. *Nature* 258:521–523.

Smith, V.J. and Ratcliffe, N.A. (1978) Host defense reactions of the shore crab, *Carcinus maenas* (L.) *in vitro. J. Marine Biol. Assoc. U.K.* 58:367–379.

Smith, V.J. and Ratcliffe, N.A. (1980a) Cellular reactions of the shore crab *Carcinus maenas*: *In vivo* hemocytic and histopathological responses to injected bacteria. *J. Invertebr. Pathol.* 35:65–74.

Smith, V.J., and Ratcliffe, N.A. (1980b) Host defense reactions of the shore crab, *Carcinus maenas* (L.): clearance and distribution of injected test particles. *J. Marine Biol. Assoc. U.K.* 60:89–102.

Smith, V.J. and Söderhäll, K. (1983) β-1,3 glucan activation of crustacean hemocytes *in vitro* and *in vivo*. *Biol. Bull.* 164:299–314.

Söderhäll, K. (1982) Prophenoloxidase activating system and melanization – a recognition mechanism of arthropods? *Dev. Comp. Immunol.* 6:601–611.

Söderhäll, K. and Cerenius, L. (1992) Crustacean immunity. *Ann. Rev. Fish Dis.* 59:2–23.

Söderhäll, K., and Smith, V.J. (1986a) Prophenoloxidase-activating cascade as a recognition and defense system in arthropods. *In:* A.P. Gupta (ed.): *Immunity in Invertebrates*, John Wiley and Sons, New York, pp 208–223.

Söderhäll, K. and Smith, V.J. (1986b) The prophenoloxidase activating system: The biochemistry of its activation and role in arthropod cellular immunity, with special reference to crustaceans. *In:* M. Brehelin (ed.): *Immunity in Invertebrates*, Springer-Verlag, Berlin, pp 208–223.

Söderhäll, K. Wingren, A., Johansson, M.W. and Bertheussen, K. (1985) The cytotoxic reaction from the freshwater crayfish, *Astacus astacus. Cell. Immunol.* 94: 326–332.

Söderhäll, K., Cerenius, L. and Johansson, M.W. (1994a) The prophenoloxidase activating system and its role in invertebrate defense, *In:* G. Beck, E.L. Cooper, G.S. Habicht and J.J. Marchalonis (eds): *Primordial Immunity: Foundation for the Vertebrate Immune System*, Academic Press, New York, pp 155–161.

Söderhäll, K. Johansson, M.W. and Cerenius, L. (1994b) Pattern recognition in invertebrates: the β-1,3-glucan binding proteins. *In:* J.A. Hoffmann, C.A. Janeway and S. Natori (eds): *Hemolytical and Humoral Immunity in Arthropods*, R.G. Landes, Austin, pp 97–104.

Sottrup-Jensen, L., Stepanik, T.M., Kristensen, T., Lonblad, P.B., Jones, C.M., Wierzbik, D.M., Magnusson, S., Domdey, H., Wetsel, R.A., Lundwall, A., Tack, B.F. and Frey, G.H. (1985) Common evolutionary origin of α_2-macroglobulin and complement components C3 and C4. *Proc. Natl. Acad. Sci. U.S.A.* 82:9–13.

Spycher, S.E., Arya, S., Iseman, D.E. and Painter, R.H. (1987) A functional, thioester-containing α_2-macroglobulin homologue isolated from the hemolymph of the American lobster *(Homarus americanus). J. Biol. Chem.* 262:14606–14611.

Stewart, J.E. and Zwicker, B.M. (1974) Induction of various vaccines for inducing resistance in the lobster *Homarus americanus. Contemp. Top. Immunobiol.* 4:233–239.

Takle, G.B. and Lackie, A.M. (1986) Chemokinetic behavior of insect haemocytes *in vitro. J. Cell Sci.* 85: 85-94.

Thornqvist, P.O., Johansson, M.W. and Söderhäll, K. (1994) Opsonic activity of cell adhesion proteins and β-1,3-glucan binding proteins from two crustaceans. *Dev. Comp. Immunol.* 18: 3–12.

Tyson, C.J. and Jenkins, C.R. (1973) The importance of opsonic factors in the removal of bacteria from the circulation of the crayfish *(Parachaeraps bicarinatus). Austr. J. Exp. Biol. Med. Sci.* 51:609–615.

Tyson, C.J. and Jenkins, C.R. (1974) The cytotoxic effect of haemocytes from the crayfish *(Parachaeraps bicarinatu)* on tumour cells of vertebrates, *Austr. J. Exp. Biol. Med. Sci.* 52: 915–923.

Vargas-Albores, F., Guzman-Murillo, A. and Ochoa, J.L. (1993a) An anticoagulant solution for haemolymph collection and prophenoloxidase studies of Panaeid shrimp *(Panaeus californiensis). Comp. Biochem. Physiol. [A]* 106:299–303.

Vargas-Albores, F., Guzman-Murillo, A. and Ochoa, J.L.A. (1993b) Lipopolysaccharide-binding agglutinin isolated from brown shrimp *(Penaeus californiensis* Holmes) haemolymph. *Comp. Biochem. Physiol. [A]* 104:407–413.

Vargas-Albores, F., Jimenez-Vega, F. and Soderhäll, K. (1996) A plasma protein isolated from brown shrimp *(Penaeus californiensis)* which enhances the activation of prophenoloxidase system by β-1,3-glucan. *Dev. Comp. Immunol.* 20:299–306.

Vasta, G.R. and Cohen, E. (1982) The specificity of *Centrunoides sculpuratus ewing* (Arizona lethal scorpion) hemolymph agglutinins. *Dev. Comp. Immunol.* 6:219–230.

Vasta, G.R. and Marchalonis, J.J. (1984) Immunobiological significance of invertebrate lectins. *Prog. Clin. Biol. Res.* 157:177–191.

Yeaton, R.W. (1981) Invertebrate lectins: Diversity of specificity, biological synthesis and function in recognition. *Dev. Comp. Immunol.* 5:535–545.

Insecta

The estimates of the numbers of insects are between 750 000 described species to several millions of undescribed ones. Generally, the insects are divided into 26 orders comprising about 1000 families (Barnes, 1987). From these immense numbers, only a few have been studied by comparative immunologists. Despite this fact, we have a relatively good knowledge of insect immune strategies. On the other hand, there is a lot of taxa of which representatives are very minute creatures, or live in extreme conditions. We have no information about their immunity. The morphofunctional endowment of these miniature animals did not allow the development of sufficiently potent immunocompetent structures and organs. There remained only a few evolutionary pathways of how to escape pathogens, such as the shortening of individual's life span, rapid change of generations, or numerous offsprings. The exoskeleton, including the cuticular lining of the foregut, hindgut and the peritropic membrane of the midgut, helped very effectively to shield them not only from desiccation and in presenting from mechanical damage, but together with the secretory activity of integumental cells also against the pathogens (Brey et al., 1993).

Blood cells

Some new information has now accumulated on the morphology and function of insect hemocytes, but their comparative classification still remains the subject of an enormous amount of controversy (Rowley and Ratcliffe, 1991; Gupta, 1991). Generally, there are described three to six or more categories of insect free blood cells more or less participating in immune reactions (Price and Ratcliffe, 1974; Ratcliffe et al., 1982; Wago, 1983; for review see Šíma and Větvička, 1990). According to a renewed classification by Gupta (1991), there are two major immunocyte types, the granulocyte or coagulocyte and the plasmocyte of which variant forms are podocytes, thrombocytes, vermicytes, and lamellocytes; and a further four other hemocyte types, the prohemocyte, the spherulocyte, the oenocytoid, and the adipohemocyte. The cells of granulocyte/coagulocyte line are round or oval and contain numerous cytoplasmic granules. The plasmatocytes are sometimes called insect macrophages because of their resemblance with vertebrate macrophages. Both these cell types represent main vectors of insect cellular immune reactions. In many insect species, fixed hemocytes are localized in various parts of the body, i.e. in corpora allata, Malphigian tubules and fat body.

Hemopoietic structures and their possible relevance to immunity

In some insects, hemocytes originate during embryogeny from a ventral strand of mesoderm from which they migrate into the epineural sinus. During later ontogeny and adultness, the hemocytes renew freely by mitotic division within the hemocoele, or in special organs which may be localized in the vicinity of fat bodies in the abdominal region or in the hypodermis (Gardiner, 1972). Some organs form only dissociated masses of accumulations of hemocytes in various stages of develoment and they could be regarded as those which have been described as "hemocyte reservoirs" for emergency purposes (Jones, 1977). Some of them could be transitional like the considered agglomerates of hemocytes in posterior parts of the body of *Musca domestica* (Nappi, 1974). Other accumulations could serve for trapping foreign material in the sense approximating the mammalian reticuloendothelial structures. It was pointed out that the pericardial cells analogously promote these functions (Poll, 1934). More developed hemopoietic organs represent the islets composed of stem cells, prohemocytes, and more differentiated cell stages, distinctly layered by a connective tissue sheath, and have been discovered in some lepidopterans and other insect species (Akai and Sato, 1971). The active proliferation within these organs was proved by using irradiation (Zachary and Hoffmann, 1973).

Humoral immunity

Insects display a wide variety of humoral responses with rather specific recognition of self from non-self. During many years of research, numerous humoral factors have been described.

A large number of different hemagglutinins were found in insect hemolymph (Ratcliffe and Rowley, 1983). Later, both anti-erythrocyte and anti-trypanososomatid agglutinins have been detected in the midgut of the tsetsefly (Molyneux et al., 1986). Although some physicochemical properties such as sugar specificity or molecular weight have been determined for natural hemagglutinins in cockroaches (Ingram et al., 1984), grasshoppers (Stebbins and Hapner, 1985), moths (Pendland and Boucias, 1986), locusts (Ingram et al., 1984) and crickets (Hapner and Jermyn, 1981), but only limited information is available for *Diptera* (McKenzie and Preston, 1992). Ingram and Molyneux (1993) compared a hemagglutinating activity in the hemolymph of three tsetse fly *Glossina* species and found that significant variations in physichochemical properties occurred between the *Glossima morsitants morsitans* and *G. palpalis gambiensis* and *G. tachinoides* agglutinins with respect to heat lability, susceptibility to ditriothreitor reduction and sensitivity to urea treatment. An N- acetyllactosamine-specific lectin called PFA has been isolated from hemolymph of a moth *Phalera flavescens*. It has molecular weight of 74 kDa and consists of two subunits. This lectin is not mitogenic for mouse cells (Umetsu et al., 1993). An agglutinin called migratorin has been isolated from the hemolymph of *Locusta migratoria*. Its molecular weight is about 650 kDa and it seems to be a polymer of eight subunits. It is heat resistant and requires presence of divalent cations (Drif and Brehelin, 1994).

The question whether the insect lectins play a role as naturally-occurring opsonins remains controversial. Some studies showed enhanced recognition and phagocytosis (Renwrantz, 1983), the other studies found no opsonization (Yeaton, 1981). One possible explanation of these discrepancies might be the presence of membrane-bound opsonins. Another possibility is the use of wrong targets. Original observation on the grasshopper agglutinin showed no opsonic activity when tested with erythrocytes, protozoan spores or bacterial cells, but when fungal blastospores were used, significant opsonic activity has been observed (Wheeler et al., 1993). The opsonic activity was blocked by galactoside, palatinose and EDTA.

Interesting properties and binding characteristics were demonstrated for a lectin derived from silkworm feces. This glycoprotein aggregates various types of mammalian cells in suspension, binds to immunoglobulins specifically at specific amino acid sequences, and recognizes the neural cell adhesion molecule (Hirayama et al., 1994). In addition, it has a strong mitogenic effect on mouse lymphocytes and increases phagocytosis of mouse peritoneal macrophages.

The beet armyworm, *Spodoptera exigua* lectin is a rather high molecular weight lectin (100–700 kDa) that binds to galactose. It is formed by two subunits of 33.2 and 34.4 kDa in equimolar concentrations. Light and electron microscopy observations localized this lectin in the granules of the granulocytes, the fat tissue being the primary site of lectin synthesis (Boucias and Pendland, 1993). Its activity is destroyed by heat and EDTA.

An unconventional approach to study insect lectins was taken by Kawauchi et al. (1993). These authors identified agglutinins in body extracts of aquatic insects by means of murine tumor cell agglutination. They found several agglutinins preferentially agglutinating sarcoma 180, Erlich sarcoma and MM-46 cells, some of them even converted an ascites tumor into a solid form. The relation of these findings to the biological role of such lectins remains to be established.

Hemocytin is an insect humoral lectin which is homologous with the mammalian von Willebrand factor. Kotani and coworkers cloned several overlapping cDNAs encoding entire silkworm lectin. The authors also introduced the cDNA into baculovirus vector. Besides von Willebrand factor, hemocytin has a homologous region with coagulation factor V and VIII (1995).

Inducible antibacterial peptides
Insects are particularly well resistant to microorganisms. Besides fast blood clotting and melanization, the synthesis of a wide array of antibacterial

peptides and polypeptides such as cecropins, attacins and defensins has been well established. This effective humoral system is residing in immune proteins that are synthesized largely in the fat body. *k*B-like motifs upstream of the immune protein genes bind the *Cecropia* immunoresponsive factor and confer high levels of inducible expression. Induction of the immune protein genes is mediated through the activation of *Cecropia* immunoresponsive factor, contingent upon thiol oxidation induced by oxidative stress (Sun and Faye, 1995).

Cecropins and defensins. Insect antibacterial peptides exhibiting a wide spectrum of activity form key defense elements in response to trauma or bacterial challenges. Well over 50 antibacterial peptides have been described in the hemolymph of immune-challenged insects (for review see Conciancich et al., 1994a; Hofmann, 1995). The reader should keep in mind that even if not every species can produce all these peptides, most of the studied species can secrete the whole range of antibacterial peptides. A sap-sucking bug *Pyrrhocoris apterus* can serve as an example: its immune blood contains a 43-residue cystein-rich anti-Gram positive bacteria peptide (belonging to defensins family), a 20-residue proline-rich peptide, and a 133-residue glycine-rich polypeptide (Cociancich et al., 1994b). The best defined families consist of defensins which act only on Gram-positive bacteria, and of cecropins, which act on Gram-positive bacteria and Gram-negative bacteria (Boman et al., 1991).

Cecropins are small 4 kDa peptides found in *Diptera* and *Lepidoptera*. These proteins were the first antibacterial peptides to be isolated and fully characterized (Hultmark et al., 1980). Cecropins attack the membrane probably via permeabilization of the lipid bilayer. So far, 18 Cecropins have been isolated (20). All of them are synthesized as larger precursors with 62–64 residues, of which 24–36 residues are absent in the mature protein.

Defensins are widely distributed among all insects (Cociancich et al., 1994a). They are small cationic 4 kDa peptides containing three domains, a flexible amino-terminal loop, central α-helix and carboxy-terminal antiparallel β-sheet. Their action is mediated through formation of voltage-dependent channels. The name defensins came from sequence similarities with mammalian defensins. They are synthesized as preprodefensins, which resembles the synthesis and maturation of some vertebrate enzymes. The conformation of defensin was determined by two-dimensional NMR and by circular dichroism study. This data show that the secondary structure of defensin A depends on the solvent nature and on the protein concentration, pH and salt presence (Maget-Dana et al., 1995).

Using a sensitive bacterial growth inhibition assay, Chalk and coworkers tested the inducible antibacterial activity in the hemolymph of mosquito *Aedes aegypti*. A peptide has been purified from immune hemolymph, and partial amino acid sequence analysis revealed substantial homology with defensins. In a subsequent study, four full-length cDNAs encoding defensin A from *A. aegypti* were cloned and sequenced. All cDNAs are 473 bases long with an open reading frame of 98 amino acids with a few substitutions in the signal peptide domain (Cho et al., 1996). The mature peptide with a predicted molecular weight of 4148 kDa, shows 80% identity and 93% similarity to *Phormia* defensin A. As with all defensins, its mRNA is produced in response to a bacterial challenge. Its level is increase 6 h after injection of bacteria, peaks at 24 h after injection and starts to decreased after 30 h. Two defensins display biological activity against Gram positive bacteria (Chalk et al., 1995).

The role of the integument in insect immunity was studied in *Bombyx mori* by epicuticular abrasion in presence of live bacteria. This type of experiment is probably more physiological than usually used to puncture the integument and inject bacteria. Antibacterial activity of cecropin was detected in the matrix of the abraded cuticle, but not in healthy cuticle. Cecropin mRNAs were detected in the underlying epithelial cells (Brey et al., 1993).

Other antibacterial factors. As both defensins and attacins were recently subjects of numerous excellent review articles, we focused our attention more on other inducible antibacterial molecules. Numerous other peptides with antibacterial properties have been described in insects. This group involves attacins studied mostly on *Hyalophora cecropia* (Hultmark, 1993; Aslin et al., 1995), sarcotoxin (Ando et al., 1987), coleoptericin (Bulet et al., 1991), hemiptericin (Conciancich et al., 1994b) and hymenoptaecin (Casteels et al., 1993).

The most prominent components of the honeybee antimicrobial defense are apideacins. Analysis of cDNA clones showed that up to 12 of these small peptides can be generated by processing of single precursor proteins. Active apidaecins are flanked by the two processing sequences, EAEPEA-EP and RR, joined together, they form a single unit that is repeated several times (Casteels-Josson et al., 1993). Each cluster is likely to be encoded by a different gene, forming a tight gene cluster. While transcriptional activation upon bacterial injection is not exceptionally fast, the multipeptide precursor nature allows for significant amplification of the response.

Diptericins are 9 kDa anti-Gram negative polypeptides originally isolated from the blood of bacterial challenged larvae of *Phormia terranovae*. They have an *N*-terminal proline-rich domain which carries a trisaccharide substitution on threonine-10 (Bulet et al., 1991).

A novel member of antibacterial peptides is lebocin. A cDNA encoding lebocin was isolated from the fat body cDNA library of *Bombyx mori* larvae. The cDNA is 844 bases long and had an open reading frame containing a signal peptide, a putative prosegment and a mature peptide. Further analysis showed that lebocin gene expression was inducible by bacterial injection, occurred tissue-specifically in the fat bodies and continued for 2 days after infection (Chowdhury et al., 1995).

Metalnikowin I is an inducible antibacterial peptide isolated from *Palomena prasina* (Chernysh et al., 1996). It is active against Gram negative bacteria and belongs to the proline-rich family of antibacterial peptides.

Hemolin is a bacteria-inducible protein belonging to the immunoglobulin superfamily with the closest similarity to the neural cell adhesion molecules (Fig. 4) (reviewed in Schmidt et al., 1993; Hofmann, 1995). It was isolated from the silk moth *Hyalophora cecropia*. It was found that hemocyte aggregation stimulated by LPS or phorbol myristate acetate was prevented by hemolin. In addition, hemolin simulated phagocytosis. Hemolin binding of bacteria is independent of bacterial LPS (Schmidt et al., 1993). It also enhanced phosphorylation of several proteins, suggesting its role in the regulation of the cellular responses via protein kinase C activation pathway (Lanz-Mendoza et al., 1996).

Readers seeking a review on principal characteristics of inducible antibacterial peptides from insects should see the excellent review by Cociancich et al. (1994a).

A qualitatively new defense factor is a complex consisting of interleukin 1 and phenol oxidase. This complex has been found in the hemolymph of the tobacco hornworm, *Manduca sexta*. It consists of prophenol oxidase, phenol oxidase and IL-1 like molecule. The entire complex has a molecular weight of 400 kDa (Beck et al., 1996).

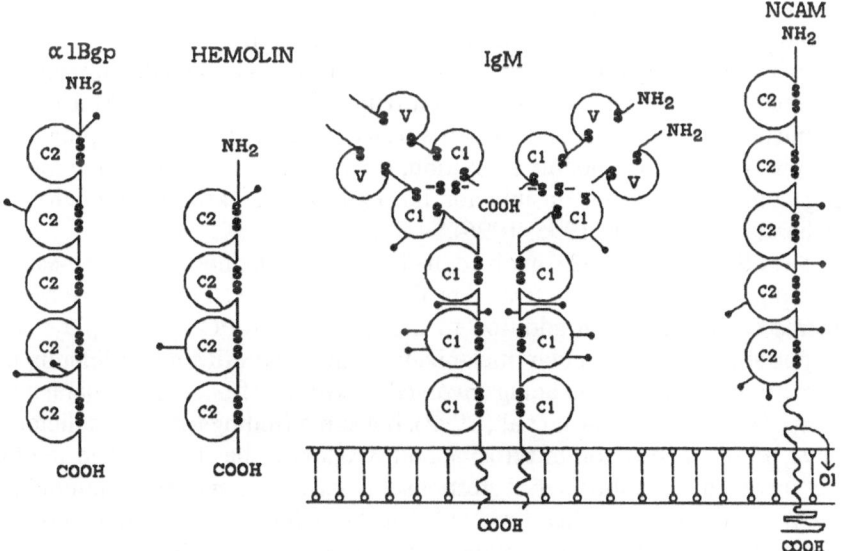

Figure 4. Schematic outline of the secondary structures of hemolin and four other members of the Ig superfamily. Different types of Ig domains V, C1, and C2 are illustrated as circles held together by disulfide bonds. The fibronectin-like domains of neuroglian are shown by open boxes. The potential *N*-linked glycosylation sites are indicated by black dots. (From Schmidt et al., 1993, with permission.)

Cellular immunology

Phagocytosis
The cockroach, *Periplaneta americana,* clears its hemocoel very rapidly of any injected material. The process of phagocytosis remains that in vertebrates – the crucial steps involve attachment, internalization and subsequent digestion. So far, no elevation of the hexose monophosphate shunt was described as the result of phagocytosis (Anderson et al., 1973). The question of which cell types are involved in phagocytosis is rather confusing and the answer probably differs from species to species. When phagocytic ability of hemocytes of *Spodoptera exigua* was compared to horse neutrophils, the *in vitro* experiments showed similar ability to internalize FITC-labeled blastospores opsonized with hemolymph lectin or horse serum, respectively. In contrast to horse neutrophils, hemocytes demonstrated no killing of fungal cells. No production of oxygen metabolites was found in insect hemocytes. Octopamine and 5-hydroxytryptamine were shown to enhance phagocytosis in cockroach hemocytes, probably via interaction with inositol triphosphate production.

It still remains to find out how insect cells actually kill bacteria. The myeloperoxidase system is not used for microbial killing (Smith and Ratcliffe, 1977). In hemolymph of *Allogamus auricollis* was found a prophenoloxidase activating cascade (proPO system) identical with that found in arthropods. This proPO system is activated by peptidoglycan, LPS and β-glucan (Brivio et al., 1996). A major non-self recognition protein of insect hemolymph is the hemocyte-derived 47 kDa protein p47. Figures 5 and 6 show the LPS-triggered spreading of insect plasmacytes and the intracellular localization of p47. Based on their data, Charalambidis and Marmaras propose that LPS-independent cell surface-associated p47 is responsible for phagocytosis and nodule formation, whereas the LPS-dependent secreting counterpart is responsible for the extracellular killing of microbes (Charalambidis et al., 1995, 1996).

The effector molecules involved in clearance are hemocyte surface-associated p47, soluble p47, activated proPO and tyrosine (for review see Marmaras et al., 1996). In addition to the defense, proPO system plays an important role in cuticular melanization and sclerotization. Melanin is sometimes suggested as being involved in killing of bacteria or parasites (Nappi et al., 1992; Nayar et al., 1992), but some findings are contradictory (Pye, 1978). The action of tyrosine and tyrosinases has been suggested to be involved in antibacterial response (Marmaras and Charalambidis, 1992). Subsequent studies revealed that the defense reaction depends on reactive tyrosine derivates involved in eumelanin biosynthesis. However, melanization and defense occur independently, as hemocytes showed high entrapment and immobilization of bacteria, but not formation of melanin (Charalambidis et al., 1994). A 47 kDa polypeptide isolated from hemocytes is also involved (Marmaras et al., 1994).

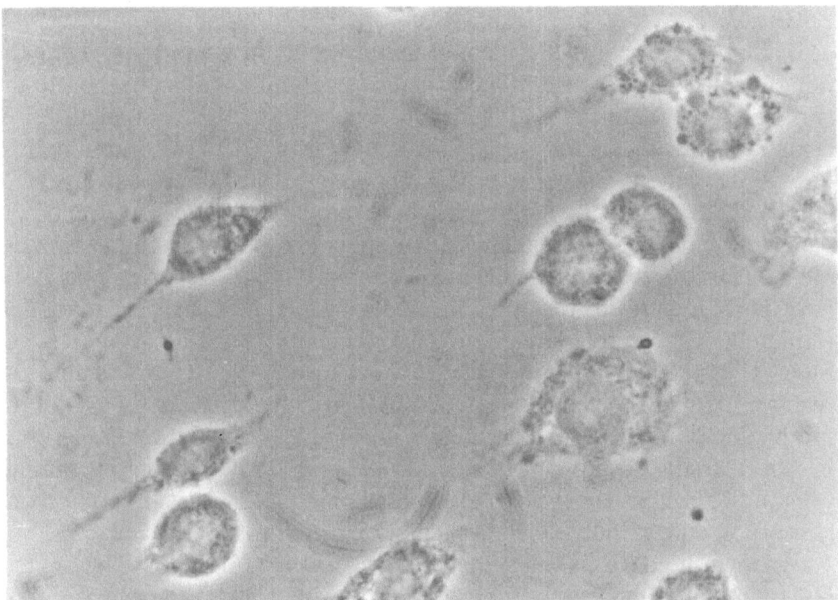

Figure 5. LPS-stimulated spreading of plasmocytes from *Ceratitis capitata*. (Reproduced courtesy of Dr. N.D. Charalambidis, Greece).

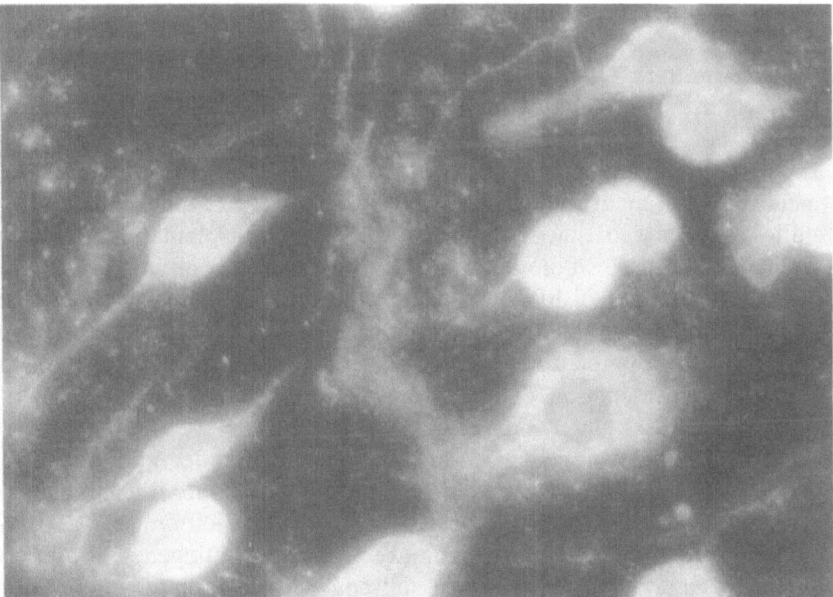

Figure 6. The same LPS-stimulated spread plasmocytes stained with anti-p47 antibody and FITC-labeled conjugate. Note the localization of p47 in vesicles. (Reproduced courtesy of Dr. N.D. Charalambidis, Greece).

Encapsulation
This type of response is typically employed against parasites. After the initial contact phase follows the lysis of the granular cells attached to the surface of the material, with subsequent clot formation and adhesion of chemoattracted plasmatocytes (Smith and Ratcliffe, 1977). A common feature is the production of a brown pigment in the innermost layers of hemocytes surrounding the invading material. The capsules enveloping the parasites were found to content both eumelanin and sclerotin (Vass et al., 1993). The cytotoxic molecules involved in killing of encapsuled parasites have been identified, but melanogenic intermediates such as quinones, semiquinones, and related quinoids have been implicated (Cox-Foster and Stehr, 1994; Miller et al., 1996), mostly because of their ability to generate reactive oxygen species. Up to recently there have been no reports of the production of reactive oxygen species during any type of immune reaction in insect. Using strains of *Drosophila melanogaster* differing in immune capabilities against the wasp parasites, Nappi et al. observed a superoxide generation during melanotic encapsulation of parasites (1995).

Endoparasitoid wasps can suppress the encapsulation response by production of immunosuppressive factor. This factor suppresses hemocyte degranulation and both encapsulation and nodule formation (Hayakawa et al., 1994).

Nodule formation is also used for elimination of large material from the internal millieu. The process remains a mixture of phagocytosis and encapsulation. The whole process is very fast and melanization starts only 5 min after initiation of clumping reaction.

Transplantation immunity
Tissue graft transplantation and the processes accompanying the rejection always represent an artificially induced situation, but for comparative immunologists it remains the only approach for testing the immunological specificity and memory. In the case of insects, despite the technical difficulties, there is a sufficient number of studies of transplantation immunity. The results of these studies are rather confusing and contradictory, mainly because of the use of different species and different grafting techniques. The experimental studies suggest that there is a relationship between the degree of transplantation reaction and taxonomical nearness of the donor/host combination (Thomas and Ratcliffe, 1982; Ratcliffe et al., 1982; Karp, 1990; Šíma and Větvička, 1990)

Karp's group developed an assay using the subcuticular epidermal layer. When *Periplaneta* recipients were xenografted with donor integument, and subsequently observed histologically for the destruction or survival of the epidermal layers, the proof of strong recognition of allogeneic grafts was obtained (George et al., 1987). In subsequent studies, integuments from *Blatta orientalis* were transplanted orthotopically onto *P. americana*. The results showed that 82% of the grafts were destroyed within 7 days after transplantation and thus demonstrated the allogeneic recognition (Karp and

Meade, 1993). Rejected and unrejected xenografts from *B. orientalis* are shown in Figures 7 and 8. However, on the basis of very different results obtained in other insect species, the question of ability of all insects to distinguish particularly allografts remains unclear.

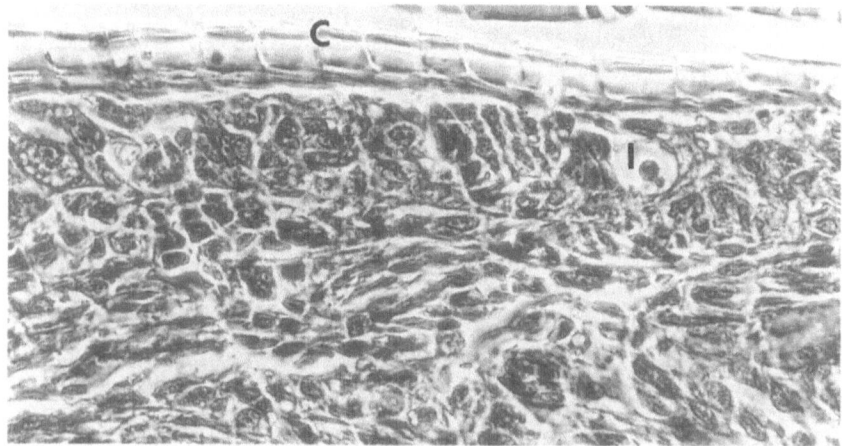

Figure 7. Rejected xenograft from *Blatta orientalis*, 7 days posttransplantation. Note the heavy cellular infiltrate (I) and total loss of the epidermal layer that is normally present under the cuticle (C). Stained with hematoxylin and eosin, magnification × 1436. (From Karp and Meade, 1993, with permission.)

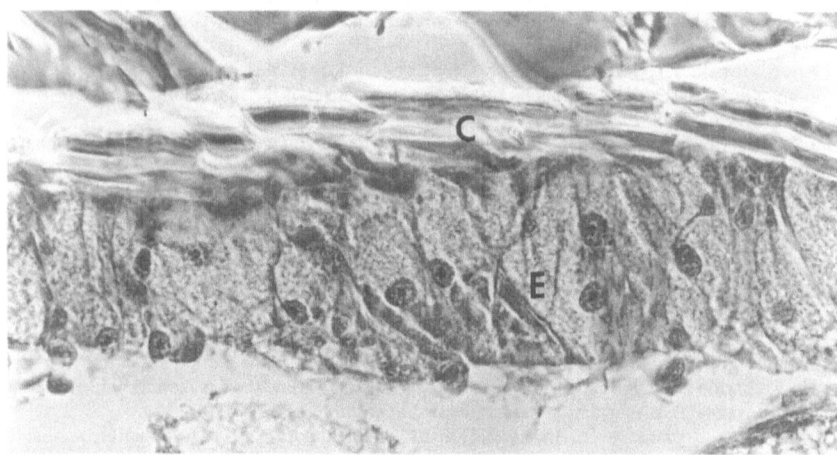

Figure 8. Unrejected xenograft from *Blatta orientalis*, 7 days posttransplantation. Note the unmistakable presence of the epidermal layer (E) under the cuticle (C) in a graft that has not been rejected. Also note the distinctive appearance of *Blatta* epidermis. Stained with hematoxylin and eosin, magnification × 1436. (From Karp and Meade, 1993, with permission.)

References

Akai, H. and Sato, S. (1971) An ultrastructural study of the haemopoietic organs of the silk-worm, *Bombyx mori. J. Insect. Physiol.* 17:1665–1676.

Anderson, R.S., Holmes, B. and Good, R.A. (1973) Comparative biochemistry of phagocytosing insect hemocytes. *Comp. Biochem. Physiol. [B]* 46: 595.

Ando, K., Okada, M. and Natori, S. (1987) Purification of sarcotoxin II from *Sarcophaga peregrina. Biochemistry* 26:226–230.

Asling, B., Dushay, M.S. and Hultmark, D. (1995) Identification of early genes in the *Drosophila* immune response by PCR-based differential display: the *Attacin A* gene and the evolution of attacin-like proteins. *Insect Biochem. Molec. Biol.* 25:511–518.

Beck, G., Cardinale, S., Wang, L., Reiner, M. and Sugumaran, M. (1996) Characterization of a defense complex consisting of interleukin 1 and phenol oxidase from the hemolymph of the tobacco hornworm, *Manduca sexta. J. Biol. Chem.* 271:11035–11038.

Boman, H.G., Faye, I., Gudmundsson, G.H., Lee, J.Y. and Lidholm, D.A. (1991) Cell-free immunity in *Cecropia*. A model system for antibacterial proteins. *Eur. J. Biochem.* 201: 23–31.

Boucias, D. and Pendland, J.C. (1993) The galactose binding lectin from the beet armyworm, *Spedoptera exigua*: distribution and site of synthesis. *Insect Biochem. Molec. Biol.* 23: 233–242.

Brey, P.T., Lee, W.J., Yamakawa, M., Koyzumi, Y., Perrot, S., Francois, M. and Ashida, M. (1993) Role of the integument in insect immunity: epicuticular abrasion and induction of cecropin synthesis in cuticular epithelial cells. *Proc. Natl. Acad. Sci. U.S.A.* 90: 6275–6279.

Brivio, M.F., Mazzei, C. and Scari, G. (1996) proPO system of *Allogamus auricollis* (Insecta): effects of various compounds on phenoloxidase activity. *Comp. Biochem. Physiol. [B]* 113: 281–287.

Bulet, P., Cociancich, S., Dimarcq, J.L., Lambert, J., Reichhart, J.M., Hoffmann, D., Hetru, C. and Hoffmann, J.A. (1991) Isolation from a coleopteran insect of a novel inducible antibacterial peptide and of new member of the insect defensin family. *J. Biol. Chem.* 266: 24520–24525.

Bulet, P., Hegy, G., Lambert, J., Van Dorsselaer, A., Hoffmann, J.A. and Hetru, C. (1995) Insect immunity. The inducible antibacterial peptide diptericin carries two o-glycans necessary for biological activity. *Biochemistry* 34:7394–7400.

Casteels, P., Ampe, C., Jacobs, F. and Tempst, P. (1993) Functional and chemical characterization of *Hymenoptaecin*, an antibacterial polypeptide that is infection-inducible in the honeybee *(Apis mellifera). J. Biol. Chem.* 268:7044–7054.

Casteels-Josson, K., Capaci, T., Casteels, P. and Tempst, P. (1993) Apidaecin multipeptide precursor structure: a putative mechanism for amplification of the insect antibacterial response. *EMBO J.* 12:1569–1578.

Chalk, R., Albuquerque, C.M.R., Ham, P.J. and Townson, H. (1995) Full sequence and characterization of two insect defensins: immune peptides from the mosquito *Aedes aegypti. Proc. R. Soc. London,* 161:217–221.

Charalambidis, N.D., Bournazos, S.N., Lambropoulou, M. and Marmaras, V.J. (1994) Defense and melanization depend on the eumalin pathway, occur independently and are controlled differentially in developing *Ceratitis capitata. Insect Biochem. Molec. Biol.* 24:655–662.

Charalambidis, N.D., Zervas, C.G., Lambropoulou, M., Katsoris, P.G. and Marmaras, V.J. (1995) Lipopolysaccharide-stimulated exocytosis of nonself recognition protein from insect hemocytes depend on protein tyrosine phosphorylation. *Eur. J. Cell Biol.* 67:32–41.

Charalambidis, N.D., Foukas, L.C. and Marmaras, V.J. (1996) Covalent association of lipopolysaccharide at the hemocyte surfaces of insects is an initial step for its internalization. Protein-tyrosine phosphorylation requirement. *Eur. J. Biochem.* 236:200–206.

Chernysh, S., Cociancich, S., Briand, J., Hetru, C. and Bulet, P. (1996) The inducible antibacterial peptides of the hemipteran insect *Palomena prasina*: Indentification of a unique family of proline-rich peptides and of a novel insect defensin. *J. Insect Physiol.* 42:81–89.

Cho, W.L., Fu, Y.C., Chen, C.C. and Ho, C.M. (1996) Cloning and characterization of cDNAs encoding the antibacterial peptide, defensin A, from the mosquito, *Aedes aegypti. Insect Biochem. Molec. Biol.* 26:395–402.

Chowdhury, S., Taniai, K., Hara, S., Kadono-Okuda, K., Kato, Y., Yamamoto, M., Xu, J., Choi, S.K., Debnath, N.C., Choi, H.K., Miyanoshita, A., Sugiyama, M., Asaoka, A. and Yamakawa, M. (1995) cDNA cloning and gene expression of lebocin, a novel member of antibacterial peptides from the silkworm, *Bombyx mori*. *Biochem. Biophys. Res. Commun.* 214:271–278.

Cociancich, S., Bulet, P., Hetru, C. and Hoffmann, J.A. (1994a) The inducible antibacterial peptides of insects. *Parasitology Today* 10:132–139.

Cociancich, S., Dupont, A., Hegy, G., Lanot, R., Holder, F., Hetru, C., Hoffmann, J.A. and Bulet, P. (1994b) Novel inducible antibacterial peptides from a hemipteran insect, the sapsucking bug *Pyrrhocori apterus*. *Biochem. J.* 300: 567–575.

Cox-Foster, D. and Stehr, J.E. (1994) Induction and localization of FAD-glucose dehydrogenase (GLD) during encapsulation of abiotic implants in *Manduca sexta* larva. *J. Insect. Physiol.* 40:235–249.

Drif, L. and Brehelin, M. (1994) Purification and characterization of an agglutinin from the hemolymph of *Locusta migratoria* (Orthoptera). *Insect Biochem. Molec. Biol.* 24: 283–289.

Gardiner, M.S. (1972) *The Biology of Invertebrates*. McGraw-Hill Book Company, New York.

George, J.F., Howcroft, T.K. and Karp, R.D. (1987) Primary integumentary allograft reactivity in the American cockroach, *Periplaneta americana*. *Transplantation* 43:514–519.

Gupta, P. (1991) Insect immunocytes and other hemocytes: roles in cellular and humoral immunity, In: A.P. Gupta (ed.): *Immunology of Insects and Other Arthropods*, CRC Press, Boca Raton, pp 19–119.

Hapner, K.D. and Jermyn, M.A. (1981) Haemagglutinin activity in the haemolymph of *Teleogryllus commodus* (Walker). *Insect Biochem.* 11:287–295.

Hayakawa, Y. (1994) Cellular immunosuppressive protein in the plasma of parasitized insect larvae. *J. Biol. Chem.* 269:14536–14540.

Hirayama, E., Ishikawa, N. and Kim, J. (1994) Further characterization of a novel lectin derived from silkworm faeces; specific binding to immunoglobulins, and activation of immunocytes. *Cell Biol. Int.* 18:257–269.

Hoffmann, J.A. (1995) Innate immunity of insects. *Curr. Opin. Immunol.* 7:4–10.

Hultmark, D. (1993) Immune reaction in *Drosophila* and other insects: a model for innate immunity. *Trends Genet.* 9:178–183.

Hultmark, D., Steiner, H., Rasmuson, T. and Boman, H.G. (1980) Insect immunity. Purification and properties of three inducible proteins from hemolymph of immunized pupae of *Hyalophora cecropia*. *Eur. J. Biochem.* 106:7–16.

Ingram, G.A., East, J. and Molyneux, D.H. (1984) Naturally occurring agglutinins against trypanosomatid flagellates in the hemolymph of insects. *Parasitology* 89:435–451.

Ingram, G.A. and Molyneux, D.H. (1993) Comparative study of haemagglutination activity in the haemolymph of three tsetse fly *Glossina* species. *Comp. Biochem. Physiol. [B]* 106: 563–573.

Jones, J.C. (1977) *The Circulatory System of Insects*. C.C. Thomas, Springfield.

Karp, R.D. (1990) Transplantation immunity in insects: Does allograft responsiveness exist? *Res. Immunol.* 11:713–725.

Karp, R.D. and Meade, C.C. (1993) Transplantation immunity in the american cockroach, *Periplaneta americana:* the rejection of integumentary grafts from *Blatta orientalis*. *Dev. Comp. Immunol.* 17:301–307.

Kawauchi, H., Hosono, M., Takayanagi, Y. and Nitta, K. (1993) Agglutinins from aquatic insects – tumor cell agglutination activity. *Experientia* 49:358–361.

Kotani, E., Yamakawa, M., Iwamoto, S., Tashiro, M., Mori, H., Sumida, M., Matsubara, F., Taniai, K., Kadono-Okuda, K. and Kato, Y. (1995) Cloning and expression of the gene of hemocytin, an insect humoral lectin which is homologous with the mammalian von Willebrand factor. *Biochem. Biophys. Acta* 1260:245–258.

Lackie, A.M. (1979) Cellular recognition of foreigness in two insect species, the American cockroach and the desert locust. *Immunology* 36:909–914.

Lanz-Mendoza, H., Bettencourt, R., Fabbri, M. and Faye, I. (1996) Regulation of the insect immune response: The effect of hemolin on cellular immune mechanisms. *Cell. Immunol.* 169:47–54.

Maget-Dana, R., Bonmatin, J.M., Hetru, C., Ptak, M. and Maurizot, J.C. (1995) The secondary structure of the insect defensin A depends on its environment. A circular dichroism study. *Biochimie* 77:240–244.

Marmaras, V.J. and Charalambidis, N.D. (1992) Certain hemocyte proteins of the medfly *Cera-titis capitata* are responsible for nonself recognition and immunobilization of *E. coli in vitro*. *Arch. Insect Biochem. Physiol.* 21:281–288.

Marmaras, V.J., Charalambidis, N.D. and Lambropoulou, M. (1994) Cellular defense mechanisms in *C. capitata*: recognition and entrapment of *E. coli* by hemocytes. *Arch. Insect Biochem. Physiol.* 26:1–14.

Marmaras, V.J., Charalambidis, N.D. and Zervas, C.G. (1996) Immune response in insects: The role of phenoloxidase in defense reaction in relation to melanization and sclerotization. *Arch. Insect Biochem. Physiol.* 31:119–133.

McKenzie, A.N.J. and Preston, T.M. (1992) Purification and characterization of a galactose-specific agglutinin from the haemolymph of the larval stages of the insect *Calliphora vomitoria*. *Dev. Comp. Immunol.* 16:31–39.

Miller, J.S., Howard, R.W., Nguven, T., Nguven, A., Rosario, R.M.T. and Staneysamuelson, W. (1996) Eicosanoids mediate nodulation responses to bacterial infections in larvae of the tenebrionid beetle *Zophobas atratus*. *J. Insect Physiol.* 42:3–12.

Molyneux, D.H., Takle, G., Ibrahim, A.B. and Ingram, D. (1986) Insect immunity to *Trypanosomatidae*. *Symp. Zool. Soc. London* 56:117–144.

Nappi, A.J. (1974) Insect hemocytes and the problems of host recognition of foreigness, *In:* E.L. Cooper (ed.): *Contemporary Topics in Immunobiology. 4. Invertebrate Immunology,* Plenum Press, New York, pp 207–224.

Nappi, A.J., Vass, E., Caron, Y. and Frey, F. (1992) Identification of 3,4-dihydroxyphenylalanine, 5,6-dihydroxyindole, and N-acetylarterenone during eumalin formation in immune reactive larvae of *Drosophila melanogaster*. *Arch. Insect Biochem. Physiol.* 20: 181–191.

Nappi, A.J., Vass, E., Frey, F. and Caron, Y. (1995) Superoxide anion generation in *Drosophila* during melanotic encapsulation of parasites. *Eur. J. Cell Biol.* 68:450–456.

Nayar, J.K., Mikarts, L.L., Knight, J.W. and Bradley, T.J. (1992) Characterization of the intracellular melanization response in *Anopheles quadrimaculatus* against subperiodic *Brugia malayi* larvae. *J. Parasitol.* 78:876–880.

Pendland, J.C. and Boucias, D. (1986) Characteristics of a galactose-binding hemagglutinin (lectin) from hemolymph of *Soidiotera exigua* larvae. *Dev. Comp. Immunol.* 10:477–487.

Poll, M. (1934) Recherches histophysiologiques sur les tubes de malphigi du *Tenebrio molitor* L. *Rec. Inst. Zool. Torley Rousseau* 5:73–126.

Price, C.D. and Ratcliffe, N.A. (1974) A reappraisal of insect haemocyte classification by the examination of blood from fifteen insect orders. *Z. Zellforsch. Mikroskop. Anat.* 147: 537–549.

Pye, A.E. (1978) Activation of prophenoloxidase and inhibition of melanization in the haemolymph of immune *Galleria mellonella* larvae. *Insect Biochem.* 8:117–123.

Ratcliffe, N.A. and Rowley, A.F. (1983) Recognition factors in insect hemolymph. *Dev. Comp. Immunol.* 7:653

Ratcliffe, N.A., White, K.N., Rowley, A.F. and Walters, J.B. (1982) Cellular defense systems of the arhropoda, *In:* N. Cohen and M.M. Sigel (eds): *The Reticuloendothelial System, Vol. 3,* Plenum Press, New York, pp 167–255.

Renwrantz, L. (1983) Involvement of agglutinins (lectins) in invertebrate defense reaction. *Dev. Comp. Immunol.* 7:603–608.

Rowley, A.F. and Ratcliffe, N.A. (1981) Insects, *In:* N.A. Ratcliffe and A.F. Rowley (eds): *Invertebrate Blood Cells Vol. 2,* Academic Press, London, pp 421–488.

Šíma, P. and Větvička, V. (1990) *Evolution of Immune Reactions.* CRC Press, Boca Raton.

Schmidt, O., Faye, I., Lindstromdinnetz, I. and Sun, S.C. (1993) Specific immune recognition of insect hemolin – review. *Dev. Comp. Immunol.* 17:195–200.

Smith, A.R. and Ratcliffe, N.A. (1977) The encapsulation of foreign tissue implants in *Galleria mellonella* larvae. *J. Invertebr. Pathol.* 23:175

Stebbins, M.R. and Hapner, K.D. (1985) Preparation and properties of haemagglutinin from haemolymoh of *Acrididae* (grasshoppers). *Insect Biochem.* 15:451–462.

Sun, S.C. and Faye, I. (1995) Transcription of immune genes in the giant silkmoth, *Hyalophora cecropia*, is augmented by H_2O_2 and diminished by thiol reagents. *Eur. J. Biochem.* 231:93–98.

Thomas, I.G. and Ratcliffe, N.A. (1982) Integumental grafting and immunorecognition in insects. *Dev. Comp. Immunol.* 9:643–654.

Umetsu, K., Yamashita, K., Suzuki, J., Yamashita, T. and Suzuki, T. (1993) Purification and characterization of an *N*-acetyllactosamine-specific lectin from larvae of a moth, *Phalera flavescens*. *Arch. Biochem. Biophys.* 301:200–205.

Vass, E., Nappi, A.J. and Carton, Y. (1993) Comparative study of immune competence and host susceptibility in *Drosophila melanogaster* parasitized by *Leptopilina boulardi* and *Asobara tabida. J. Parasitol.* 79:106–112.

Wago, H. (1983) Cellular recognition of foreign materials by *Bombyx mori* phagocytes: II. Role of hemolymph and phagocyte filopodia in the cellular reactions. *Dev. Comp. Immunol.* 7: 199–208.

Wheeler, M.B., Stuart, G.S. and Hapner, K.D. (1993) Agglutinin mediated opsonization of fungal blastospores in *Melanoplus differentialis* (Insecta). *J. Insect Physiol.* 39:477–483.

Yeaton, R.W. (1981) Invertebrate lectins: Diversity of specificity, biological synthesis and function in recognition. *Dev. Comp. Immunol.* 5:535–545.

Zachary, D. and Hoffmann, J.A. (1973) The haemocytes of *Calliphora erythrocephala* (Meig) (Diptera). *Z. Zellforsch. Mikroskop. Anat.* 141:55–73.

Mollusca

The molluscs represent an ancient group of animals whose ancestral representatives resembled the segmented annelid worms and were bilaterally symmetric (Runneggar and Pojeta, 1985). The secondary modifications of the molluscan body pattern probably occurred in the Precambrian,
as all of the major molluscan groups were established before the end of
the Cambrian (Conway Morris, 1993). The members of the phylum are
divided into seven classes, the *Gastropoda*, composed of marine, freshwater, and terrestrial animals known as limpets, periwinkles, cowries,
whelks, cone shells, sea slugs, sea butterflies, snails, and slugs, the *Mono-
placophora* and the *Polyplacophora* (chitons), considered to be the most
primitive groups, the *Aplacophora,* the strange worm-like molluscs commonly called solenogasters, the *Bivalvia*, including common forms such as
clams, oysters, and mussels, the *Scaphopoda*, the tusk or tooth shells, and
finally the *Cephalopoda*, comprising well known nautili, cuttlefish, squids,
and octopuses.

The molluscs are certainly among the most successful assemblages of
animals. No malacologist knows exactly how many species are alive, and
estimates run from 45 000 to 50 000 (Barnes, 1987; May, 1990) to 150 000
(Meglitsch, 1967) species worldwide. The fundamental parts of the molluscan body, the head, visceral mass, and foot may be more or less modified or reduced. The process of torsion and unequal growth of one or more
of these body parts gave rise to a tremendous plasticity of the molluscan
basic body pattern, which consequently led to the emergence of a great
variety of types. The specific features of molluscs differentiating them
from any higher coelomate metazoans are the absence of segmentation and
the massive reduction of the coelom (only to pericardium, gonads, and
excretory organs in adultness). This alteration is connected with a transformation of primary body cavity into vascular system, the haemocoel. In all
molluscs, the circulatory system is open, with the exception of cephalopods, where it reached a maximal specialization complexity never found
in any invertebrate taxa (Schipp, 1987). The other specificity of molluscs
is a mantle. Its epidermis secretes material for calcareous shell formation.
On the basis of a large spatial reduction of the coelom and the mesenchymal body organization of molluscs, some authors render their body construction as to be rather acoelomate (von Salvini-Plaven, 1988), and deny
their kinship as a sister group to coelomate echiurans, sipunculids, and
annelids (Scheltema, 1993; Kim et al., 1996). Understanding the molluscan origins and relationships together with their uniqueness in the body
construction might help us to explain their proper immune strategy more
precisely.

The hemopoietic tissues and organs

Very few data are known with respect to both hemopoietic structures and immune reactions in taxa of molluscs other than gastropods, bivalves and cephalopods. In polyplacophorans, only free amoebocytes resembling those of bivalves and gastropods capable of endocytosis and fixed phago-cytes with a high immune recognition abilities have been described, the later placed in ctenidia, foot and digestive gland (Cuénot, 1914; Crichton et al., 1973; Kilby et al., 1973).

The gastropods

In gastropods, hemopoiesis takes place freely in hemolymph or throughout the body like in *Lymnea stagnalis* (Sminia, 1974), in epithelia and connec-tive tissue of the walls of blood sinuses, and in the region of heart-kidney and mantle like in *Bulinus* species (Kinoti, 1971), and ventricle and lung cavity (Sminia, 1981). In the snail *Biomphalaria glabrata*, a main source of hemocytes (even sometimes considered as amoebocyte-producing organ) is the connective tissue between mantle region and pericardium (Lie et al., 1975, Lie and Heyneman, 1976). Small clusters of amoeboblasts and secondary cells often seen in the mitotic division have been detected in a loose reticulum of anterior, dorsal and posterior portions of the mantle walls (Jeong et al., 1983). The organ became enlarged during the process of encapsulation.

Up to the present, there is no consensus in unified classification of gastropod cell categories. Some authors recognize only one basic blood cell type, the amoebocyte (Sminia, 1972), others admit the presence of two types, the granulocyte and the hyalinocyte (Harris, 1975; Yoshino, 1976; Krupa et al., 1977). Both fundamental cell types are actively phagocytic cells, the former seem to be young cells maintaining their proliferative capacity, the latter are the terminal differentiated cells. The morphology of these cells and their developmental stages resemble those seen in mono-cyte-macrophage lineage of vertebrates. It is clear that between these cell categories, a row of intermediate stages may exist (for review see Sminia, 1981). It was shown that the gastropod hemocytes can divide even if they are engaged in phagocytosis and encapsulation (Sminia, 1974). This means that they may serve as progenitor cells giving rise to daughter cells. A spe-cialized population forming the pool of stem cells was also suggested in some species (Brown and Brown, 1965; Kinoti, 1971; Lie et al., 1975). The number of circulating hemocytes may increase rapidly under some cir-cumstances such as rising temperature, wounding or irritation (Sminia, 1972; Stumpf and Gilbertson, 1978). As the increase is not accompanied by mitotic divisions, a release of cells into circulation from the connective tissue cell pool has been proposed (Sminia, 1974). Thus, the gastropod hemocytes represent a renewing population which in case of necessity could be mobilized from other parts of the body. It resembles an inflammation process in mice, where new macrophages are released from the blood and Peyer's patches.

In the connective tissue of the digestive gland region and around blood vessels of snail *Helix pomatia* (Reade, 1968; Bayne, 1973b) and *L. stagnalis* (Sminia et al., 1979b), the fixed phagocytes have been revealed. The third cell type capable of endocytosis represents the pore cells. These cells closely resembling the podocytes and nephrocytes of crustaceans were observed throughout the connective tissue (Sminia, 1972; Boer and Sminia, 1976). Despite the fact that the main role of pore cells is the synthesis and secretion of blood pigments, they actively endocytose soluble protein material and even very small particles. According to the majority of authors, they serve in ultrafiltration rather than in immune functions (Fletcher and Cooper Willis, 1982).

Hemocytes from *Haliotis tuberculata* responded to addition of vertebrate growth factors such as epidermal growth factor or insulin by DNA and protein syntheses (Lebel et al., 1996). The establishment of *in vitro* primary cultures of hemocytes described in this study provides a suitable model for further studies of the properties of molluscan cells.

The bivalves

A vast number of different types of free blood cells have been described in bivalves, but direct proof for hematopoiesis is still lacking. Similarly to gastropods, the bivalve hemocytes are believed to arise from the connective tissue (Moore and Lowe, 1977). Because of variable morphology and unclear origin of bivalve blood cells, the proposals for unifying classification of them are still under critical analyses. A practical and useful categorization has been suggested first by Cuénot (1897) and Takatsuki (1934), than by Moore and Lowe (1977) and recently by Cheng (1981). Cheng surveyed all data available up to that time and generalized a new scheme using three fundamental types of cells common to all bivalves, the granulocytes, the hyalinocytes, and the serous cells. It is noteworthy to mention that at least in some bivalves the serous cells are capable of phagocytosis of small particles. They are produced by so-called Keber's glands or pericardial glands forming part of the excretory system (Keber, 1851).

The cephalopods

The highly developed circulatory system of cephalopods is unique in its endowment of a specialized vessel system by incomplete endothelia and which is not found in any other invertebrate animal. The large vessels have generally three-layered walls like those of vertebrates (Kawaguti, 1970; Schipp, 1987). The distinct, well-organized blood-forming tissue called white bodies or Hensen's gland appears as a multilobed structure in the orbital pit area behind the eyes (Stuart, 1968). Inside these organs, the hemocytoblasts, various developmental cell stages (primary and secondary leukoblasts), and transitional cells could be seen. The new cells leave the organ via haemal sinusoids (Cowden and Curtis, 1981). These blood-forming structures are similar in all cephalopod species that have been studied.

It seems that in cephalopods only one free blood cell type exists. It could be compared to both vertebrate monocytes and granulocytes (Cowden and Curtis, 1981). Practically, no significant phagocytic power of these cells has been detected (Bayne, 1973a). The primitive "RES" has been described in octopus, *Eledone cirrosa,* comprising phagocytes in the gills, the stroma of posterior salivary gland, and white bodies (Stuart, 1968).

Humoral immunity

The molluscs have probably been investigated for the presence of agglutinins more than any other invertebrate group. (For excellent reviews of these studies, see Pemberton, 1974; Gold and Balding, 1975; Ey and Jenkin, 1982; Vasta and Marchalonis, 1987; Šíma and Větvička, 1990; Vasta et al., 1994.) Considerable evidence has been presented that these substances are important for the identification of foreign material by hemocytes (Sminia et al., 1979a) and that the titers of these molecules are influenced by bacterial or parasite challenge (Loker et al., 1994).

Similar to other species, various lectins have been discovered in gastropods, but their true physiological function remains unknown. The majority of agglutinins are weak and often of indeterminate specificity (for review see Pemberton, 1974). Pemberton studied 134 species and found hemagglutinins in 59% of tested species. Infection of *B. glabrata* with trematode *Echinostoma paraensei* induces production of unique polypeptides designated as Group 1 and Group 2 molecules which can bound to the surface of various non-self objects such as bacteria, erythrocytes or larvae in a lectin-like fashion (Hertel et al., 1994). The antibacterial substances have only rarely been found in snails (Kamiya et al., 1984). These substances have rather high molecular weight and are often composed of subunits. A glycoprotein termed achacin has been isolated and characterized from the mucus of giant African snail, *Achatina fulica*. Achacin kills both gram-positive and negative bacteria, but has no bacteriolytical activity (Osuka-Fuchino et al., 1992).

A wide range of immunoreactive vertebrate bioactive peptides has been found in hemocytes of the fresh water snail *Viviparus ater*, including bombesin, calcitonin, CCK-8, insulin, oxytocin, serotonin, somatostatin, vaso-pressin, and substance P (Ottaviani et al., 1992). These data support the hypothesis of a common origin of the immune and neuroendocrine systems. In this light the distinction between hormones and neurotransmit-ters becomes merely semantic, as the same substance may belong to one or more systems. The addition of corticotropin-releasing factor to hemocytes induced the release of epinephrine and dopamine, which is a phenomenon being considered as an ancestral type of stress response (Ottaviani et al., 1994). On the other hand, preincubation of hemocytes with IL-2 or anti-IL-2 antibodies eliminated the corticotropin-releasing factor-induced release of amines. The authors suggest the existence of an ancestral recep-tor on molluscan hemocytes, capable of binding both IL-2 and corticotro-pin-releasing factor (Ottaviani et al., 1994). To further confirm this hypo-thesis, hemocytes of two molluscs, *V. ater* and *Planobarius corneus* were preincubated with IL-1α, IL-1β, TNF-α and TNF-β before addition of corticotropin-releasing factor. The biological response of these hemocytes was strongy reduced (Ottaviani et al., 1995a). On the other hand, mollus-can hemocytes are responsive to mammalian neuropeptides and their frag-ments (Genedani et al., 1994).

The bivalve molluscs posses similar types of nonspecific humoral im-munity as other invertebrate creatures. Various humoral factors include agglutinins (Tyler, 1946) and bactericidins (Mori et al., 1980) have been identified. *Crassostrea gigas* hemolymph contain lectins gigalin H that agglutinates human erythrocytes and gigalin E that agglutinates horse erythrocytes. *In vivo* challenge by bacteria resulted in an increased hemolymph lectin titer compared to unchallenged animals (Olafsen, 1995). Gigalins also act as opsonins in uptake of bacteria by oyster hemocytes.

American oyster, *C. virginica* hemagglutinins were characterized by Tripp (1974b). When rabbit erythrocytes were used as targets, hemagglu-tinins acted as opsonins (Tripp, 1966). However, the question of what role these proteins might play in infection remained unanswered.

The hemolymph of the hard clam, *Mercenaria mercenaria*, contains a lectin that agglutinates bacteria and enhances the phagocytosis of only that particular bacteria (Arimoto and Tripp, 1977). On the other hand, hemocytes of *M. mercenaria* do phagocytose a variety of prey even in the absence of hemolymph (Tripp, 1992a). The importance of opsonins for defense mechanisms of this animal is thus in question. Further studies con-firmed that lectins in the *M. mercenaria* hemolymph agglutinate red blood cells, bacteria, and yeast, but are not involved in enhancement of immune reactions (Tripp, 1992b). These lectins thus represent one end of a spec-trum of molluscan lectin interaction ranging from total dependence (Prowse and Tait, 1969) through various degrees of enhanced internaliza-tion (Sminia et al., 1979a) to total independence.

The albumen glands in the sea hare, *Dolabella auricularia*, contain several proteins with antineoplastic, antibacterial and antifungal properties

(Iijima et al., 1994). Some of them were found to inhibit growth of human lymphoma cell lines (Beckwith et al., 1993). Another defense factor named Aplysianin-A was isolated from the albumen gland of *Aplysia kurodai*. It is an antibacterial glycoprotein with a 50% amino acid sequence homology to achacin, an antibacterial glycoprotein of the giant African snail *Achatina fulica* (Takamatsu et al., 1995). An interesting protein has been isolated from the hemolymph of *B. glabrata*. This plasma protein binds to a variety of vertebrate immunoglobulins (Hahn et al., 1996), but its physiological role is unknown.

Cellular immunity

In an attempt to elucidate the plasma components involved in prey recognition by *B. glabrata* hemocytes, Fryer and Bayne used positively and negatively charged synthetic beads coated with homologous or heterologous plasma (1966). The results of these experiments suggested that the phagocytosis is modulated in a strain-specific manner by absorbed plasma components.

Encapsulation and cytotoxic killing of metazoan parasites in molluscs appear to depend on the effective hemocyte binding to the surface of invading parasite, and the cytotoxic activation of the cell via membrane transduction of the binding signal (van der Knaap and Loker, 1990). These questions have been studied using *Schistotoma mansoni* susceptible and resistant snails. Hemocytes from resistant snail, *B. glabrata* (Bayne et al., 1980) and *L. stagnalis* (Dikkeboom et al., 1988) bind directly to the surface of *S. mansoni* larvae leading subsequently to the killing of encapsulated larvae. On the other hand, hemocytes of the susceptible snails bind to the parasite, but do not kill them (Bayne et al., 1980). To address the hypothesis that the hemocytes of resistant and susceptible snails differ in their surface receptors, Coustau and Yoshino compared surface polypeptides on the membrane of circulating hemocytes from resistant and susceptible snails (1994). A 66-kDa polypeptide present on the surface of both adherent and nonadherent hemocytes from the susceptible strains, but only weakly or even absent on cells from the resistant snails, was the only difference found so far. In addition, exposure of these molluscs to *S. mansoni* enhanced the expression of this polypeptide in susceptible snails.

Hemocytes of the hard clam *M. mercenaria* migrate toward secreted bacterial products by chemotaxis (Fawcett and Tripp, 1994). This finding is potentially important in solving the question of how leukocytes accumulate at sites of injury.

Several lysosomal enzymes have been identified in molluscan hemocytes, among which are lysozyme and β-glucuronidase (Cheng et al., 1975) and acid phosphatase (Yoshino and Chang, 1976). The role of these enzymes is probably in destroying and digesting foreign material. The oxidative killing mechanisms similar to those known in vertebrates were found in some molluscs, such as the scallop *Patinopecten yessoensis* (Nakamura et al., 1985) or blue mussel *M. edulis* (Noel et al., 1993). On the other hand, these mechanisms were not found in several clams tested (Lopez-Gomez et al., 1994; Torreilles et al., 1996).

More detailed study of the hemocytes of the pond snail, *L. stagnalis,* showed that these cells generate reactive oxygen intermediates upon contact with foreign material (Adema et al., 1993). Based on inhibition of NAPDH-oxidase by several inhibitors, the authors conclude that the source of reactive oxygen intermediates is NAPDH-oxidase activity.

The phagocytosis of foreign material in molluscs is well described (for review see Tripp, 1994a, Bayne and Fryer, 1994). The fate of phagocytosed prey is determined by its nature and localization. Inert material is excreted to the surrounding millieu, whereas the readily digestible materials are degraded by phagocytes. Numerous studies showed that the phagocytosis and/or bacterial clearance are influenced by the bacterial or parasite challenge (Nunez et al., 1994).

Hematopoietic neoplasia is a commercially important blood cell disease of the soft-shelled clam *Mya arenaria*. Diseased cells differ morphologically from their normal counterparts. Comparative study in normal and diseased hemocytes showed that yeast cells adhered only to the surface of normal hemocytes (Beckmann et al., 1992). The tests of phagocytic activity revealed that diseased cells not only do not bind the prey, but are also unable to ingest the material.

Using an anti-nitric oxide synthase (NOS) antibody, the molluscan hemocytes were found to be positive for NOS. Stimulation of animals with LPS resulted in increased NOS reactivity (Franchini et al., 1995). Biochemical studies revealed that the enzyme is 30% membrane bound and 70% cytoplasmic. In addition, mollusc hemocytes produce nitric oxide which is a bactericide substance. The production of nitric oxide was found to be related to phagocytosis (Ottaviani et al., 1993). Its presence in this group of animals is further proof that some critical molecules have been conserved over the evolution.

The tissue rejection has been studied in sufficient details. Recently, a molluscan heart as a new model for use in transplantation studies has been introduced. Direct measurement of heartbeat in addition to the more classical histological observation can be made, making evalutation of graft rejection more accurate. When allografts were implanted heterotopically into the hemocoel of the snail *B. glabrata*, most hearts continued to beat for the whole duration of the study (i.e., 6 months). Histological changes included temporary encapsulation by hemocytes, formation of fluid-filled sac around the graft and decreased organization of the myocardium (Sullivan et al., 1995). Another study showed no differences in fates among different xenografts or between xenografts and allografts for up to 180 days (Sullivan et al., 1992, 1993). These observations of prolonged survival differ from previously published studies by other investigators, most of whom have described rapid graft rejection (Cheng and Galloway, 1970). In contrast to their previous findings, Sullivan's group showed a fast (1 to 15 days) destruction of heart xenografts from seven different genera implanted into the hemocoel of *B. glabrata*. Hearts from *Helisoma trivolis* were the only exception, surviving for at least 6 months (Sullivan et al., 1995). The detailed morphological observations of heterotopic heart transplants from this study are in Figures 9–20.

When amoebocyte-producing organ was transplanted into the hemocoel of the snails of different strains, no allograft rejection was observed even 151 days after transplantation. Moreover, some transplants even showed mitotic figures (Sullivan, 1990).

Immunoreactive cytokines appear to play a role in the linkage of the immune and nervous system in molluscs. *M. edulis*, a marine bivalve, reacts to the vertebrate IL-1, IL-6 and TNF. These cells respond to these cytokines, both *in vivo* and *in vitro,* in a manner similar to that of human granulocytes (Hughes et al., 1990). As in human system, the effects of IL-1 are, at least partially, the result of its induction of TNF formation. In addition, endogenous immunoreactive IL-1, IL-6 and TNF were found in *M. edulis* hemolymph, immunocytes and pedal ganglia (Hughes et al., 1992). IL-1α, IL-1β and TNF-α strongly stimulate hemocyte motility, phagocytic activity and production of nitric oxide synthase (Ottaviani et al., 1995b). Immunocytochemical study (Fig. 21) found the presence of TNF-α both in hemocytes isolated from *B. glabrata* (Ouwe-Missi-Oukem-Boyer et al., 1994). In addition, TNF-α reactivity was demonstrated also in the hemolymph and digestive gland (Fig. 22). The activity of IL-1 differed between *Schistosoma*-resistant and susceptible snails (Granath et al., 1994). Based on the data of results describing the action of various cytokines in molluscs, we can summarize that they are important ancestral molecules which maintained basic functions throughout the evolutionary development.

Some strains of *B. glabrata* posses a soluble plasma factor stimulating hemocytes to kill larval schistosomes. This factor was shown to be an IL-1-like molecule (Granath et al., 1994). The *Schistosoma*-resistant snails maintained significantly higher level IL-1-like factor than *Schistosoma*-susceptible strains of snails. Furthermore, the levels of this molecule were increased after exposure to *S. mansoni*.

Nonadherent hemocytes from *Planorbarius corneus* possess NK-like activity, as they can lyse K 562 cells (Franceschi et al., 1991a). This activity is reduced after 18 h incubation, but can be preserved by addition of human IL-2. None of the other cytokines tested (IL-4, IFNγ, TNF-α and TNF-β) was able to produce similar effect (Franceschi et al., 1991b). Surprisingly, hemocytes with NK-like activity bound several monoclonal antibodies directed against mammalian NK cells and adhesion molecules. The adherent and nonadherent cells differed in this respect, the adherent

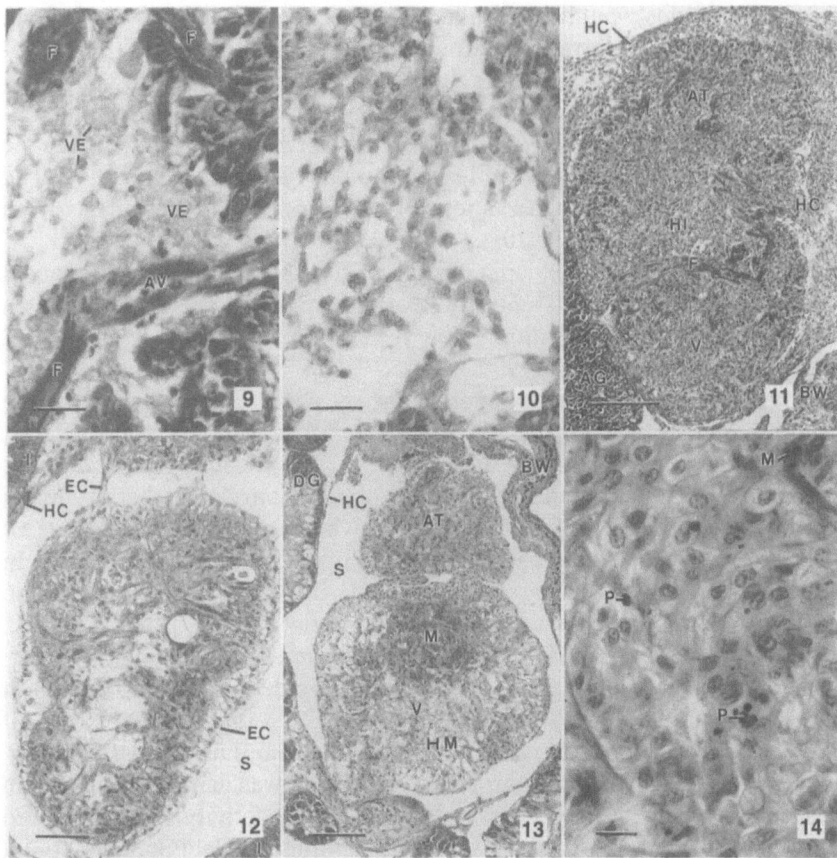

Figures 9–14. Hearts of *Helisoma trivolis* implanted as xenografts in *B. glabrata*. Histological sections. Figure 9. Ventricle of control heart showing cytoplasmic extensions that, in section, resemble vesicles (VE) in the lumen. Atrioventricular valve (AV); fascicle of myofibers (F). Scale bar = 20 μm. Figure 10. Ventricular lumen at day 1, showing infiltrating hemocytes and absence of cytoplasmic extensions seen in control hearts. Scale bar = 20 μm. Figure 11. Xenograft at 7 days, showing extensive encapsulation and infiltration by hemocytes. Atrium (AT); recipient albumin gland (AG); recipient body wall (BW); fascicle of myofibers (F); hemocytic capsule (HC); hemocytic infiltrate (HI). Scale bar = 100 μm. Figure 12. Ventricle at 30 days. Note hypertrophied epicardium (EC), which is attached in places to the hemocytic capsule (HC); fluid-filled space (S) separating heart from surrounding hemocytic capsule; and absence of infiltrating hemocytes. Recipient intestine (I). Scale bar = 50 μm. Figure 13. Xenograft at 120 days, showing degeneration of myocardium of ventricle. Note extremely thin hemocytic capsule (HC). Atrium (AT); recipient body wall (BW); recipient digestive gland (DG); hypochromatic, disorganized myocardium (HM); myofibers (M); fluid-filled space separating heart from hemocytic capsule (S); ventricle (V). Scale bar = 100 μm. Figure 14. Polygonal cells in myocardium at 30 days. Myofiber (M); pigment (P). Scale bar = 10 μm. (From Sullivan et al., 1995, with permission.)

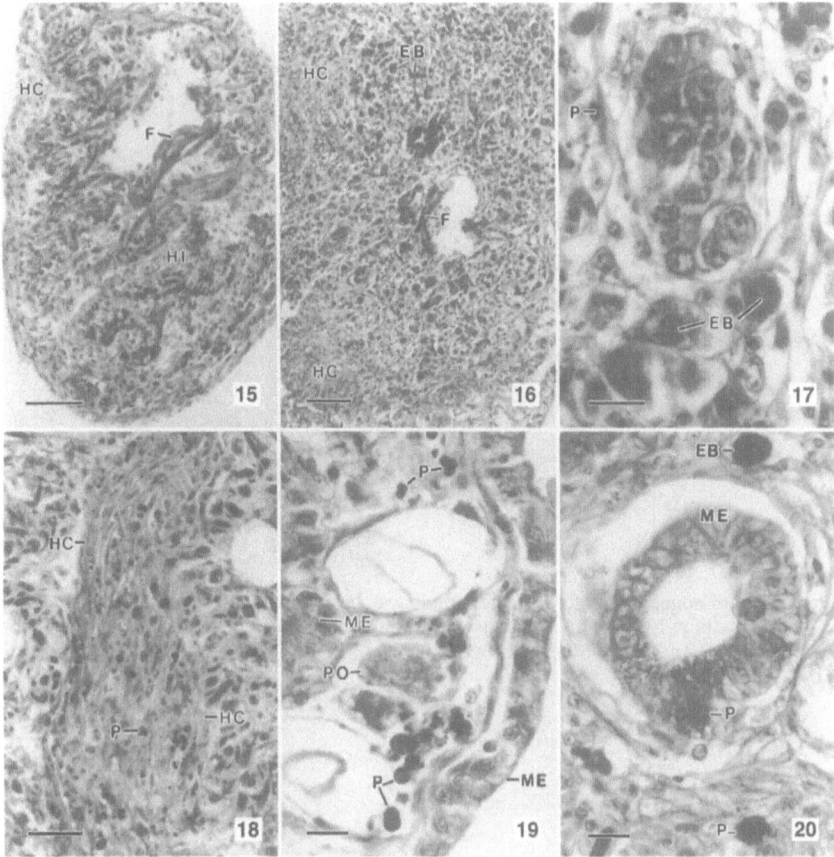

Figures 15–20. Hearts of *Planorbis atticus* (Figs 15–16) and *Physa virgata* (Fig. 20) implanted as xenografts in *B. glabrata*. Histological sections. Figure 15. Ventricle of *P. atticus* at day 1. Note hemocytic capsule (HC) and hemocytic infiltrate (HI). Fascicle of myofibers (F). Scale bar = 50 µm. Figure 16. Xenograft at 3 days, showing extensive necrosis and lack of clearly defined border. Eosinophilic body (EB); fascicle of myofibers (F); hemocytic capsule (HC). Scale bar = 50 µm. Figure 17. Multinucleate cell in xenograft at 3 days. Eosinophilic body (EB); pigment (P). Scale bar = 10 µm. Figure 18. Xenograft in recipient's pseudobranch at 7 days, showing complete necrosis and replacement by hemocytes. Hemocytic capsule (HC); pigment (P). Scale bar = 30 µm. Figure 19. Pigment within hemocytes in connective tissue of rectal ridge in a xenograft recipient at 30 days. Mantle epithelium (ME); pigment (P); pore cells (PO). Scale bar = 10 µm. Figure 20. Vesicle of pigmented mantle epithelium (ME) surrounded by a hemocytic capsule at 7 days. Eosinophilic body (EB); pigment (P). Scale bar = 10 µm. (From Sullivan et al., 1995, with permission.)

hemocytes were positive for anti-CD5, anti-CD34, anti-CD45RA, anti-CD54, anti-CD61 and anti-CD71 antibodies. Both populations were stained with anti-CD1a, anti-CD16, anti-CD26, anti-CD29 and anti-CD56 antibodies. The possible cross-reactivity instead of real specificity cannot be ruled out, but the amount of positive antibodies is rather too large for this simple explanation. These findings support the hypothesis of the very early

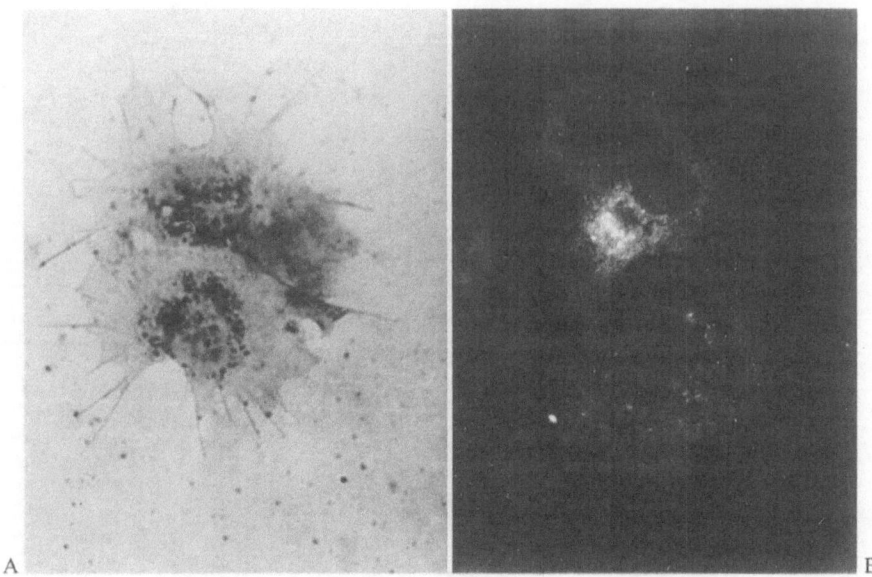

Figure 21. Immunoperoxidase (A) and immunofluorescence (B) staining of *B. glabrata* hemocytes with polyclonal anti-human TNF-α antibodies. Positive staining was observed in cell granules. (From Ouwe-Missi-Oukem-Boyer et al., 1994, with permission.)

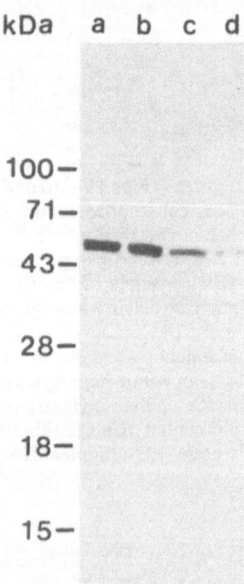

Figure 22. Identification of TNF-α immunoreactive components in hemocytes (a), hemolymph (b), digestive gland (c) and whole snail extract (d). Western blot analysis of different mollusc fractions (50 µg protein) was performed using rabbit polyclonal anti-human TNF-α antibodies. (From Ouwe-Missi-Oukem-Boyer et al., 1994, with permission.)

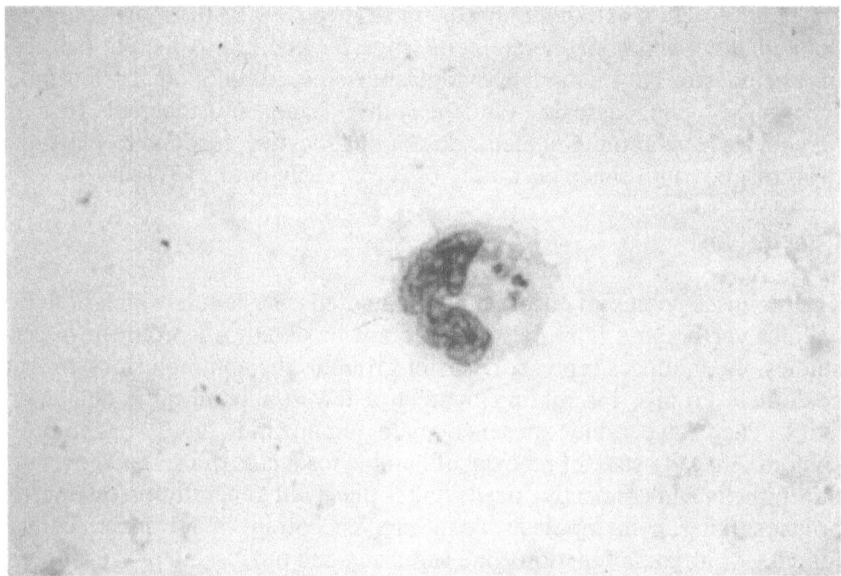

Figure 23. Morphology of the adherent (spreading) hemocytes stained with May-Grünwald-Giemsa. (From Franceschi et al., 1991, with permission.)

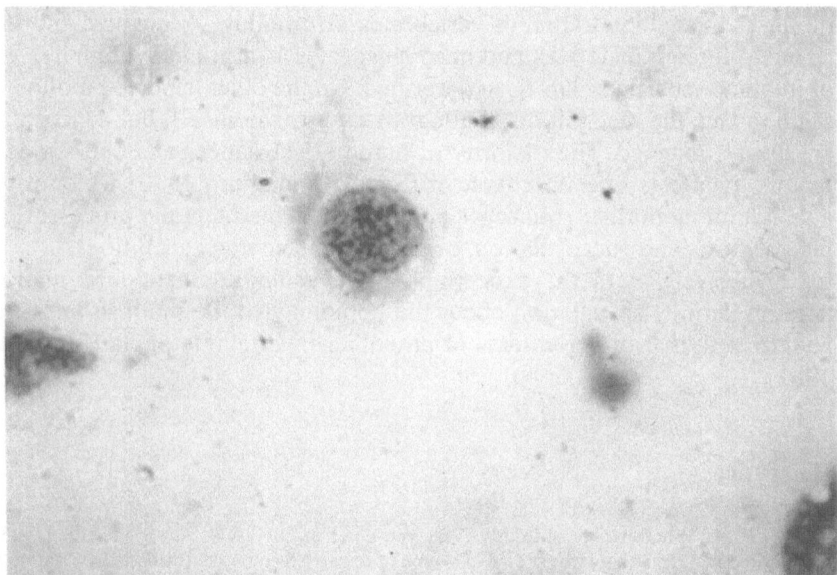

Figure 24. Morphology of the nonadherent (round) hemocytes stained with May-Grünwald-Giemsa. (From Franceschi et al., 1991, with permission.)

appearance of NK activity. Subsequent studies showed the morphology of both adherent and nonadherent cells (Figs. 23 and 24). Adherent (spreading) hemocytes have abundant cytoplasm rich in pseudopodia. Their main function is phagocytosis. On the other hand, nonadherent (round) hemocytes have a round nucleus, do not phagocytize, respond to PHA and form rosettes with sheep erythrocytes (Franceschi et al., 1991b).

Conclusions

The molluscs posses an efficient internal defense system in which both the cellular vectors and humoral effectors are involved. According to recent studies, the molluscs represent a sister group to the common stock of true coelomate groups, resembling them in a few synapomorphic characteristics. They have rather mesenchymate organization with very limited coelom. A steady state of renewal of hemocytes inside the vascular system, the epithelia and connective tissue, takes place but generally, no distinctive organ serving hematopoiesis (with the exception of advanced cephalopods), or immune functions like capturing and processing of the antigen, i.e., no equivalents of vertebrate lymph node or spleen are present. The molluscan hemocytes exert some macrophage-like properties, produce a number of humoral defense substances with strong lytic, cytotoxic, agglutinating, and opsonizing properties. Moreover, the presence of cytokine-like molecules (IL-1, IL-2, IL-6, TNF-α, and TNF-β) has been observed in molluscan hemocytes (Franceschi et al., 1994). The existence of two effective phagocytic systems, the free cells and fixed cells, approximate the molluscs in analogy to that of vertebrates. All molluscs recognize and respond to foreign materials, and in some species a surprisingly high degree of immune specificity has been described. On the other hand, the molluscs seem to lack the recognition of allograft tissue as non-self, but xenografts are always rejected. The majority of humoral substances and cell-surface factors manifests a lectin character (Suzuki and Mori, 1990) which may attach to or agglutinate microbes and parasites, facilitate the processes of phagocytosis and encapsulation, or possess opsonizing capabilities.

Conclusively, with the exception of the analogies mentioned above, nothing permits speculation about the homology of the mollusc immune pattern with that of vertebrates or any other invertebrate phyla (Fletcher, 1982).

References

Adema, C.M., van Deutekom-Mulder, E.C., van der Knaap, W.P.W. and Sminia, T. (1993) NADPH-oxidase activity: the probable source of reactive oxygen intermediate generation in hemocytes of the gastropod *Lymnea stagnalis*. *J. Leukocyte Biol.* 54:379–383.
Arimoto, R. and Tripp, M.R. (1977) Characterization of a bacterial agglutinin in the hemolymph of the hard clam, *Mercenaria mercenaria*. *J. Invertebr. Pathol.* 30:406–413.

Barnes, R.D. (1987) *Invertebrate Zoology.* Saunders College Publ., Philadelphia.

Bayne, C.J. (1973a) Internal defense mechanisms of *Octopus dolfeini. Malacol. Rev.* 6: 13–17.

Bayne, C.J. (1973b) Molluscan internal defense mechanism: the fate of C^{14}-labelled bacteria in the land snail *Helix pomatia* (L.). *J. Comp. Physiol.* 86:17–25.

Bayne, C.J. (1983) Molluscan immunobiology. *In:* A.S.M. Saleuddin and K.M. Wibur (eds): *The Mollusca 5, Physiology, Part 2,* Academic Press, New York, pp 407–486.

Bayne, C.J. and Fryer, S.E. (1994) Phagocytosis and invertebrate opsonins in relation to parasitism. *In:* G. Beck, E.L. Cooper, G.S. Habicht and J.J. Marchalonis (eds): *Primordial Immunity: Foundation for the Vertebrate Immune System,* Ann. New York Acad. Science, New York, pp 162–177.

Bayne, C.J., Buckley, P.M. and Dewan, P.C. (1980) Macrophage-like hemocytes of resistant *Biomphalaria glabrata* are cytotoxic for sporocysts of *Schistosoma mansoni in vitro. J. Parasitol.* 29:131–142.

Beckmann, N., Morse, M.P. and Moore, C.M. (1992) Comparative study of phagocytosis in normal and diseased hemocytes of the bivalve mollusc *Mya arenia. J. Invertebr. Pathol.* 59: 124–132.

Beckwith, M., Urba, W.J. and Longo, D.L. (1993) Growth inhibition of human lymphoma cell lines by marine products, dolastatins 10 and 15. *J. Natl. Cancer Inst.* 85:483–488.

Boer, H.H. and Sminia, T. (1976) Sieve structure of slit diaphragms of podocytes and pore cells of gastropod molluscs. *Cell Tissue Res.* 170:221–229.

Brown, A.C. and Brown, R.J. (1965) The fate of thorium dioxide injected into the pedal sinus of *Bullia* (Gastropoda: Prosobranchia). *J. Exp. Biol.* 42:509–519.

Cheng, T.C. (1981) Bivalves, *In:* N.A. Ratcliffe and A.F. Rowley (eds): *Invertebrate Blood Cells, Vol. 2,* Academic Press, London, pp 233–300.

Cheng, T.C. and Galloway, P.C. (1970) Transplantation immunity in molluscs: The histoincompatibility of *Helisoma duryi normale* with allografts and xenografts. *J. Invertebr. Pathol.* 15: 177–192.

Cheng, T.C., Rodrick, G.E., Foley, D.A. and Koehler, S.A. (1975) Release of lysozyme from hemolymph cells of *Mercenaria mercenaria* during phagocytosis. *J. Invertebr. Pathol.* 25: 261–265.

Conway Morris, S. (1993) The fossil record and the early evolution of the matazoa. *Nature* 361: 219–225.

Coustau, C. and Yoshino, T.P. (1994) Surface membrane polypeptides associated with hemocytes from *Schistosoma mansoni*-susceptible and -resistant strains of *Biomphalaria glabrata* (Gastropoda). *J. Invertebr. Pathol.* 63:82–89.

Cowden, R.R. and Curtis, S.K. (1981) Cephalopods, *In:* N.A. Ratcliffe and A.F. Rowley (eds): *Invertebrate Blood Cells, Vol.2,* Academic Press, London, pp 301–323.

Crichton, R. and Lafferty, K.J. (1975) The discriminatory capacity of phagocytic cells in the chiton (*Liolophura gaimardi*), *In:* W.H. Hildemann and A.A. Benedict (eds): *Immunologic Phylogeny,* Plenum Press, New York, pp 89–98.

Crichton, R., Killby, V.A.A. and Lafferty, K.J. (1973) The distribution and morphology of phagocytic cells in the chiton *Liolophura gaimardi. Austr. J. Exp. Biol. Med. Sci.* 51: 357–372.

Cuénot, L. (1897) Les globules sanguins et les organes lymphoides des invertébrés (Revue critique et nouvelles recherches). *Arch. Anat. Microsc. Morphol. Exp.* 1:153–192.

Cuénot, L. (1914) Les organes phagocyaires des mollusques. *Arch. Zool. Exp. Gen.* 54: 267–305.

Dikkeboom, R., Bayne, C.J., van der Knaap, W.P.W. and Tijnagel, J.M.G.H. (1988) Possible role of reactive forms of oxygen in *in vitro* killing of *Schistosoma mansoni* sporocysts by hemocytes of *Lymnea stagnalis. Parasitol. Res.* 75:148–154.

Ey, P.L. and Jenkin, C.R. (1982) Molecular basis of self/non-self discrimination in the invertebrata. *In:* N. Cohen and M.M. Sigel (eds): *The Reticuloendothelial System, Vol. 3,* Plenum Press, New York, pp 321–391.

Fawcett, L.B. and Tripp, M.R. (1994) Chemotaxis of *Mercenaria mercenaria* hemocytes to bacteria *in vitro. J. Invertebr. Pathol.* 63: 275-284.

Fletcher, T.C. and Cooper-Willis, C.A. (1982) Cellular defense systems of the mollusca, *In:* N. Cohen and M.M. Sigel (eds): *The Reticuloendothelial System, Vol. 3,* Plenum Press, New York, pp 141–166.

Franceschi, C., Cossarizza, A., Monti, D. and Ottaviani, E. (1991a) Cytotoxicity and immunocyte markers in cells from the freshwater snail *Planorbarius corneus* (L.) (Gastropoda: Pulmonata): implications for the evolution of natural killer cells. *Eur. J. Immunol.* 21:489–493.

Franceschi, C., Cossarizza, A., Ortolani, C., Monti, D. and Ottaviani, E. (1991b) Natural cytotoxicity in a freshwater pulmonate mollusc: an unorthodox comparative approach. *Adv.
Neuroimmunol.* 1:99–113.

Franchini, A., Conte, A. and Ottaviani, E. (1995) Nitric oxide: an ancestral immunocyte effector molecule. *Adv. Neuroimmunol.* 5:463–478.

Fryer, S.E. and Bayne, C.J. (1966) Phagocytosis of latex beads by *Biomphalaria glabrata*
hemocytes is modulated in a strain-specific manner by adsorbed plasma components. *Dev.
Comp. Immunol.* 20:23–37.

Genedani, S., Bernardi, M., Ottaviani, E., Franceschi, C., Leung, M.K. and Stefano, G.B.
(1994) Differential modulation of invertebrate hemocyte motility by CRF, ACTH, and its
fragments. *Peptides* 15:203–206.

Gold, E.R. and Balding, P. (1975) *Receptor-Specific Proteins: Plant and Animal Lectins*.
Excerpta Medica, Amsterdam.

Granath, W.O., Connors, V.A. and Tarleton, R.L. (1994) Interleukin 1 activity in haemolymph
from strains of the snail *Biomphalaria glabrata* varying in susceptibility to the human blood
fluke, *Schistosoma mansoni*: presence, differential expression, and biological function. *Cytokine* 6:21–27.

Hahn, U.K., Fryer, S.E. and Bayne, C.J. (1996) An invertebrate (molluscan) plasma protein that
binds to vertebrate immunoglobulins and its potential for yielding false-positives in anti-
body-based detection systems. *Dev. Comp. Immunol.* 20:39–50.

Harris, K.R. (1975) The fine structure of encapsulation in *Biomphalaria glabrata. Ann. N.Y.
Acad. Sci.* 266:446–464.

Hertel, L.A., Stricker, S.A., Monroy, F.P., Wilson, W.D. and Loker, E.S. (1994) *Biomphalaria
glabrata* hemolymph lectins: binding to bacteria, mammalian erythrocytes, and to sporocysts
and rediae of *Echinostoma paraensei. J. Invertebr. Pathol.* 64:52–61.

Hughes, T.K., Smith, E.M., Chin, R., Cadet, P., Sinisterra, J., Leung, M.K., Shipp, M.A., Scharrer, B. and Stefano, G.B. (1990) Interaction of immunoactive monokines (interleukin 1 and
tumor necrosis factor) in the bivalve mollusc *Mytilus edulis. Proc. Natl. Acad. Sci. U.S.A.* 87:
4426–4429.

Hughes, T.K., Smith, E.M., Leung, M.K. and Stefano, G.B. (1992) Immunoactive cytokines in
Mytilus edulis nervous and immune interactions. *Acta Biol. Hung.* 43:269–273.

Iijima, R., Kisugi, J. and Yamayaki, M. (1994) Biopolymers from marine invertebrates. XIV.
Antifungal property of dolabellanin A, a putative seld-defense molecule of the sea hare,
Dolabella auricularia. Biol. Pharm. Bull. 17:1144–1146.

Jeong, K.H., Lie, K.J. and Heyneman, D. (1983) The ultrastructure of the amebocyte-producing
organ in *Biomphalaria glabrata. Dev. Comp. Immunol.* 7:217–228.

Kamiya, H., Muramoto, K. and Ogata, K. (1984) Antibacterial activity in the egg mass of a sea
hare. *Experientia* 40:947–949.

Kawaguti, S. (1970) Electron microscopy of muscle fibres in blood vessels and capillaries of
cephalopods. *Biol. J. Okayama Univ.* 16:19–28.

Keber, G.A.F. (1851) *Beiträge für Anatomie und Physiologie der Weichtiere*. Gebr. Bornträger,
Königsberg.

Killby, V.A.A., Crichton, R. and Lafferty, K.J. (1973) Fine structure of the phagocytic cells in
the chiton, *Liolophura gaimardi. Austr. J. Exp. Biol. Med. Sci.* 51:373–391.

Kim, C.B., Moon, S.Y., Gelder, S.R. and Kim, W. (1996) Phylogenetic relationships of annelids, molluscs, and arthropods evidences from molecules and morphology. *J. Morph. Evol.*
43:207–215.

Kinoti, G.K. (1971) Observations on the infection of bulinid snails with *Schistosoma mattheei.
Parasitology* 62:161–170.

Krupa, P.L., Lewis, L.M. and Del Vecchio, P. (1977) *Schistosoma haematobium* in *Bulinus guer-
nei*: electron misroscopy of hemocyte-sporocyst interactions. *J. Invertebrate Pathol.* 30:35–45.

Lebel, J.M., Giard, W., Favrel, P. and Boucaud-Camou, E. (1996) Effects of different vertebrate growth factors on primary cultures of hemocytes from the gastropod molluscs, *Haliotis
tuberculata. Biol. Cell* 86:67–72.

Lie, K.J., Heyneman, D. and You, P. (1975) The origin of amoebocytes in *Biomphalaria
glabrata. J. Parasitol.* 61:574–576.

Lie, K.J. and Heyneman, D. (1976) Studies of resistance in snails. 6. Escape of *Echinostome lin-
doense* sporocysts from encapsulation in the snail heart and subsequent loss of the host's ability to resist infection by the same parasite. *J. Parasitol.* 62:298–302.

Loker, E.S., Couch, L. and Herten, L.A. (1994) Elevated agglutination titres in plasma of *Biomphalaria glabrata* exposed to *Echinostoma paraensei* – characterization and functional relevance of a trematode-induced response. *Parasitology* 108:17–26.

Lopez-Gomez, C., Villalba, A. and Bachere, E. (1994) Absence of generation of active oxygen radicals by the hemocytes of the clam, *Ruditapes decussatus* (Mollusca: Bivalvia) coupled with phagocytosis. *J. Invertebr. Pathol.* 64:188–192.

May, R.T. (1990) How many species? *Phil. Trans. R. Soc. London B* 330:293–304.

Meglitsch, P.A. (1967) *Invertebrate Zoology.* Oxford Univ. Press, London.

Moore, M.N. and Lowe, D.M. (1977) The cytology and cytochemistry of the hemocytes of *Mytilus edulis* and their responses to experimental' injected carbon particles. *J. Invert. Pathol.* 29:18–30.

Mori, K., Tone, Y., Suzuki, T., Kasahara, K. and Nomura, T. (1980) Defense mechanisms of molluscs, I. Bactericidal and agglutinin activities in the scallop tissues. *Bull. Jap. Soc. Fish* 46:717–723.

Nakamura, N., Mori, K., Inooka, S. and Nomura, T. (1985) *In vitro* production of hydrogen peroxide by the amoebocytes of the scallop, *Patinopecten yessoensis* (Jay). *Dev. Comp. Immunol.* 9:407–417.

Noel, D., Bachere, E. and Mialhe, E. (1993) Phagocytosis associated chemiluminescence of hemocytes in *Mytilus edulis (Bivalvia). Dev. Comp. Immunol.* 17:483–493.

Nunez, P.E., Adema, C.M. and Dejongbrink, M. (1994) Modulation of the bacterial clearance activity of haemocytes from the freshwater mollusc, *Lymnaea stagnalis,* by the avian schistosome, *Trichobilharzia ocellata. Parasitology* 109:299–310.

Olafsen, J.A. (1995) Role of lectins (C-reactive protein) in defense of marine bivalves against bacteria. *Adv. Exp. Med. Biol.* 371:343–348.

Otsuka-Fuchino, H., Watanabe, Y., Hirakawa, C., Tamiya, T., Matsumoto, J.J. and Tsuchiya, T. (1992) Bactericidal action of a glycoprotein from the body surface mucus of gian african snail. *Comp. Biochem. Physiol. [C]* 101:607–613.

Ottaviani, E., Franchini, A. and Fontanili, P. (1992) The presence of immunoreactive vertebrate bioactive peptide substances in hemocytes of the freshwater snail *Viviparus ater* (Gastropoda, Prosobranchia). *Cell. Mol. Neurobiol.* 12:455–462.

Ottaviani, E., Paemen, L., Cadet, P. and Stefano, G.B. (1993) Evidence for nitric oxide production and utilization as a bacteriocidal agent by invertebrate immunocytes. *Eur. J. Pharmacol.* 248:319–324.

Ottaviani, E., Franchini, A., Caselgrandi, E., Cossarizza, A. and Franceschi, C. (1994) Relationship between corticotropin-releasing factor and interleukin-2: evolutionary evidence. *FEBS Lett.* 351:19–21.

Ottaviani, E., Caselgrandi, E. and Franceschi, C. (1995a) Cytokines and evolution: *in vitro* effects of IL-1α, IL-1β, TNF-α and TNF-β on an ancestral type of stress response. *Biochem. Biophys. Res. Comm.* 207:288–292.

Ottaviani, E., Franchini, A., Cassanelli, S. and Genedani, S. (1995b) Cytokines and invertebrate immune responses. *Biol. Cell* 85:87–91.

Ouwe-Missi-Oukem-Boyer, O., Porchet, M., Capron, A. and Dissous, C. (1994) Characterization of immunoreactive TNF alpha molecules in the gastropod *Biomphalaria glabrata. Dev. Comp. Immunol.* 18:211–218.

Pemberton, R.T., (1974) Anti-A and anti-B of gastropod origin. *Ann. N.Y. Acad. Sci.* 234:95–107.

Prowse, R.H. and Tait, N.N. (1969) *In vitro* phagocytosis by amoebocytes from the hemolymph of *Helix aspersa* (Muller). I. Evidence for opsonic factor(s) in serum. *Immunology* 17:437–443.

Reade, P.C. (1968) Phagocytosis in invertebrates. *Austr. J. Exp. Biol. Med. Sci.* 46:219–229.

Runnegar, B. and Pojeta, J. (1985) Origin and diversification of the Mollusca, In: E.R. Trueman and M.R. Clarke (eds): *The Mollusca, Vol. 10, Evolution,* Academic Press, Orlando, pp 1–57.

Scheltema, A.H. (1993) Aplacophora as progenetic aculiferans and the coelomate origin of molluscs as the sister taxon of sipuncula. *Biol. Bull.* 184:57–78.

Schipp, R. (1987) The blood vessels of cephalopods. A comparative and functional survey. *Experientia* 43:52–537.

Síma, P. and Větvička, V. (1990) *Evolution of immune reactions.* CRC Press, Boca Raton.

Sminia, T. (1972) Structure and function of blood and connective tissue cells of the fresh water pulmonate *Lymnea stagnalis* studied by electron microscopy and enzyme histochemistry. *Z. Zellforsch. Mikrosk. Anat.* 130:497–526.

Sminia, T. (1974) Haematopoiesis in the freshwater snail *Lymnea stagnalis* studies by electron microscopy and autoradiography. *Cell Tiss. Res.* 150:443–454.

Sminia, T. (1981) Gastropods, *In:* N.A. Ratcliffe and A.F. Rowley (eds): *Invertebrate Blood Cells, Vol. 1,* Academic Press, London, pp 191–232.

Sminia, T., van der Knaap, W.P.W. and Edelenbosch, P. (1979a) The role of serum factors in phagocytosis of foreign particles by blood cells of the fresh water snail *Lymnea stagnalis. Dev. Comp. Immunol.* 3:37–44.

Sminia, T., van der Knaap, W.P.W. and Kroese, F.G.M. (1979b) Fixed phagocytes in the freshwater snail *Lymnea stagnalis. Cell. Tissue Res.* 196:545–548.

Sullivan, J.T. (1990) Long-term survival of heterotopic allografts of the amoebocyte-producing organ in *Biomphalaria glabrata* (Mollusca: Pulmonata). *Trans. Am. Microsc. Soc.* 109: 52–60.

Sullivan, J.T., Andrews, J.A. and Currie, R.T. (1992) Heterotopic heart transplant in *Biophalaria glabrata* (Mollusca: Pulmonata): fate of allografts. *Trans. Am. Microsc. Soc.* 111:1–15.

Sullivan, J.T., Weir, G.O. and Brammer, S.R. (1993) Heterotopic heart transplants in *Biomphalaria glabrata* (Mollusca: Pulmonata). Fate of congeneric xenografts. *Dev. Comp. Immunol.* 17:467–474.

Sullivan, J.T., Brammer, S.R., Hargraves, C.D. and Owens, B.S. (1995) Heterotopic heart transplants in *Biomphalaria glabrata (Mollusca: Pulmonata)*: Fate of xenografts from seven pulmonate genera. *Invertebrate Biol.* 114:151–160.

Stuart, A.E. (1968) The reticulo-endothelial apparatus of the lesser octopus *Eledone cirrosa. J. Pathol. Bacteriol.* 96:401–412.

Stumpf, J.L. and Gilbertson, D.E. (1980) Differential leucocytic responses of *Biomphalaria glabrata* to infection with *Schistosoma mansoni, J. Invertebrate Pathol.* 36:217–218.

Takamatsu, N., Skiba, T., Muramoto, K. and Kamiva, H. (1995) Molecular cloning of the defense factor in the albumen gland of the sea hare *Aplysia kurodai. FEBS Lett.* 377:373–376.

Takatsuki, S.I. (1934) On the nature and functions of the amoebocyte of *Otrea edulis. Quart. J. Microsc. Sci.* 76:379–431.

Torreilles, J., Guerin, M.C. and Roch, P. (1996) Reactive oxygen species and defense mechanisms in marine bivalves. *Compt. Rendus Acad. Sci.* 319:209–218.

Tripp, M.R. (1966) Hemagglutinin in the blood of the oyster hemolymph. *J. Invertebr. Pathol.* 8:478–484.

Tripp, M.R. (1974a) Molluscan immunity. *Ann. N.Y. Acad. Sci.* 234:23–27.

Tripp, M.R. (1974b) Oyster hemolymph proteins. *Ann. N.Y. Acad. Sci.* 234:18–20.

Tripp, M.R. (1992a) Phagocytosis by hemocytes of the hard clam, *Mercenaria mercenaria. J. Invertebr. Pathol.* 59:478–484.

Tripp, M.R. (1992b) Agglutinins in the hemolymph of the hard clam, *Mercenaria mercenaria. J. Invertebr. Pathol.* 59:228–234.

Tyler, A. (1946) Natural heteroagglutinins in the body fluids and seminal fluids of various invertebrates. *Biol. Bull.* 90:213–219.

van der Knaap, W.P.W. and Loker, E.S. (1990) Immune mechanisms in trematode-snail interactions. *Parasitol. Today* 6:175–182.

Vasta, G.R. and Marchalonis, J.J. (1987) Invertebrate agglutinins and the evolution of humoral and cellular recognition factors. *In:* A.H. Greenberg (ed.): *Invertebrate Models. Cell Receptors and Cell Communication,* Karger, Basel, pp 134–150.

Vasta, G.R., Ahmed, H., Fink, N.E., Elola, M.T., Marsh, A.G., Snowden, A. and Odom, E.W. (1994) Animal Lectins as self/non-self recognition molecules Biochemical and genetic approaches to understanding their biological roles and evolution. *In:* G. Beck, E.L. Cooper, G.S. Habicht and J.J. Marchalonis (eds): *Primordial Immunity: Foundation for the Vertebrate Immune System* Ann. New York Acad. Science, New York, pp 55–73.

von Salvini-Plaven, L. (1988) Annelida and mollusca – a prospectus, *In:* W. Westheide and C.O. Hermans (eds): *The Ultrastructure of Polychaeta,* Gustav Fischer Verlag, Stuttgart, pp 383–396.

Yoshino, T.P. (1976) Encapsulation response of the marine prosobranch *Cerithidea californica* to natural infections of *Renicola buchanani* sporocysts (Trematoda: Renicolidae). *Int. J. Parasitol.* 6:423–431.

Yoshino, T.P. and Cheng, T.C. (1976) Fine structural localization of acid phosphatase in granulocytes of the pelecypod *Mercenaria mercenaria. Trans. Am. Microsc. Soc.* 95:215–220.

Deuterostomes

Echinodermata

Members of the phylum *Echinodermata* (around 6000) are the only known species of deuterostome invertebrates. The living echinoderms are at present classified into three subphyla, the *Crinozoa*, the *Asterozoa*, and the *Echinozoa* which comprise the echinoderms commonly known as sea lilies and feather stars, then starfish, sea stars, basket stars, serpent stars and brittle stars, and finally sea urchins, heart urchins, sand dollars, and sea cucumbers (Moore, 1966–1978). They are exclusively marine, largely bottom dwellers. Their most striking feature is pentamerous radial symmetry of the body pattern secondarily derived from a bilateral ancestor.

All echinoderms differ from other invertebrate phyla by the presence of a calcareous internal skeleton, and unique and highly specialized coelomic canals and surface appendages forming the water vascular system, and tube feet. The entrocoelic origin of the coelom is a further characteristic distinguishing echinoderms from other advanced invertebrate taxa such as annelids, molluscs, and arthropods and giving them more kinship with protochordates and chordates. This is stressed by the radial and indeterminate cleavage, the basic characteristics of all deuterostomes (Tab. 1).

The coelomic derivatives of echinoderms and their possible role in immunity

During embryogeny the two lateral pouches separating from the archenteron give rise to a coelomic cavity which develops further by subdivision into coelomic vesicles forming the axocoel, the hydrocoel, and the somatocoel, all corresponding to the protocoel, the mesocoel, and the metacoel cavities of other deuterostomes. The cells composing archenteral pouch walls become mesoderm (Clark, 1964). In the adultstage, the coelom is

Table 1. Basic features common to echinoderms and chordates

Radial indeterminate cleavage
Deuterostomy (blastopore = anus)
Invaginative origin of nervous system
Enterocoelic coelom
Internal mesodermal skeleton
Pluteus larva

---

specifically arranged in the representatives of different echinoderm subtaxa into two major isolated parts forming the tubular coelomic system
(including the water vascular, hemal, and perihemal systems), and the perivisceral coelom surrounding the gut and internal organs (hyponeural sinus;
Millott, 1967; Nichols, 1969).

The immune significance of coelomic tubular and perivisceral systems

The water vascular, hemal, and perihemal systems

The water vascular system function is to maintain a sufficient hydrostatic
pressure within the hydraulic organ, the tube feet, serving for animal locomotion, burrowing, and feeding. Its canals are lined with a ciliated epithelium. Some special bulbous structures or organs associated with the circum
oral ring part of the water vascular system, the spongy bodies, the Tiedemann's bodies, and the Polian vesicles, have supposed immune significance.
The coelomocyte-like cells found in inner epithelial folds of Tiedemann's
bodies may be considered to be precursors of coelomocytes. Because of the
position of these organs where both circulatory systems meet, they may function in capturing antigens (analogs of lymph nodes; Karp and Coffaro, 1982).
The so-called axial organ or gland is situated in close proximity to the stone
canal which is a part of the water vascular system that communicates with the
external environment, and may even be of a higher immunological importance (see below). A part of the sea water enters the madreporite, passing
through the Tiedemann's bodies into the perivisceral coelom (Ferguson,
1984), so these bodies may function as the sources of coelomic fluid.

The hemal system is composed of a number of ring channels often surrounded by separate extensions of the perihemal spaces and sinuses. From
the oral ring of the hemal system, a vessel running along the stone canal
ascends through the axial organ which could be, in this sense, considered
as a derivative of hemal system. Sea cucumbers possess the most developed
hemal system. Hemal vessels may be composed of muscle and connective
tissue layers with interior endothelium and exterior ciliated peritoneum
(Herreid et al., 1976). The function of the hemal system is supposed to be
internal transport and distribution of food materials.

The perihemal system, in the species where it is present, follows and
mimics the canals or lacunas of a hemal system. It surrounds both the axial
organ and the stone canal, connecting with them near the madreporite. It probably has the analogical functions of transport similar to the hemal system.

Perivisceral coelom

The perivisceral coelom is filled by a fluid which cannot be homologized
either with invertebrate hemolymph or vertebrate blood. The coelomic
cavity is lined by ciliated endothelium propels which the circulation of the

coelomic fluid. Several coelomocyte types produced by coelomic perito-neum could be distinguished moving freely in the coelomic fluid or also pervading body tissues. They are active in phagocytosis, the removal and transport of particulate waste to the gills, podia, or axial organ. In holothu-rians they carry particulate waste and nitrogenous material in crystalline form to the gonads, the respiratory tree, and the intestine where waste accu-mulates and is extruded from the body (Smith, 1981). The coelomocytes take part in forming a clot in the case of tissue damage (see below).

Other structures engaged in the immunity
Body wall. Among many epidermal cell types, the mucous cell could be of some immune importance. Together with macrodetritus, the plankton and potentially pathogenic microorganisms are trapped in the mucus and then swept away by epidermal ciliated cells (with the exception of holothurians). In some species the mucus containing plankton is carried by ciliary cur-rents to the mouth (auxiliary ciliary feeding).

Humoral immunity

Hemagglutinins have been extensively studied in echinodermates. Various lectins with ability to agglutinate chicken, rabbit, calf, pig, mouse and human erythrocytes have been described for review see (Ey and Jenkin, 1982; Šíma and Větvička, 1990). Lectin isolated from coelomic fluid of the sea urchin *Paracentrotus lividus* agglutinates not only rabbit and human erythrocytes, but also promotes adhesion of autologous coelomocytes (Canicatti et al., 1992). The lectin did not discriminate between blood groups from the human ABH system, which is consistent with previous findings in other echinoderms (Parrinello et al., 1976) or in other inverte-brates (Hall and Rowlands, 1974; Anderson and Good, 1975). This activity is Ca^{2+} but not Mg^{2+} dependent and is heat-stable, therefore it looks like this agglutinin could be a C-type lectin. More detailed characterization reveal-ed that this agglutinin has MW over 200 kDa. Under reducing conditions, three distinct bands of MW 174 kDa, 137 kDa and 76 kDa, respectively, were observed. Similar to those findings, a 220 kDa lectin released by the coelomocytes of the sea cucumber *Holothuria polii* agglutinates amoebo-cytes and spherulocytes (Canicatto and Rizzo, 1991). The authors suggest-ed the possibility that a cell attachment sequence responsible for cell adhe-sion, originally described for fibronectin (Pierschbacher and Ruoslahti, 1984), might occur in *Paracentrotus lividus* agglutinin. The finding in the amino acid sequence of the lectin isolated from *Anthocidaris crassispina* (Giga et al., 1987) of a fibronectin-like tripeptide sequence seems to sup-port this possibility.

Based on these data, one may conclude that the lectins of echinoderms have, similar to those of other phyla, a double function: to defend the organism against invading bacteria or altered cells, and to play a fundamental role as cell adhesion molecules involved in cell-cell and cell-matrix interaction. The hypothesis suggesting the primary role of these molecules in primitive self/non-self recognition remains to be proved.

Coelomocytes of *H. polii* produce two trypsin-resistant lytic proteins. One of them is Ca^{2+} dependent, heat-sensitive and is present in all animals, whereas the second one is Ca^{2+} independent, heat-stable and occurs only after antigenic stimulation (Canicatti and Ciulla, 1987). The lytic action of these hemolysins is not related to the complement system (Canicatti et al., 1987). An interesting hemolysin was described in *S. droebachiensis*. It also acted as an opsonin, which resembles effects of complement fragments and suggested the possibility of alternative pathway type of complement-like system in echinoderms (Bertheussen and Seljelid, 1982; Bertheussen, 1983). Ten years later, a hemolysin with similar properties was discovered in the coelomic fluid from *Strongylocentrotus nudus* (Ito et al., 1992a, b). Further characterization showed that the hemolysin has a molecular weight of 200 kDa (Osada et al., 1993). The hemolytic activity of this molecule was inhibited by mammalian complement inhibitors, suggesting that it might have similar functions to those of the mammalian complement (Bertheussen, 1983; Ito et al., 1992a). These findings suggest that some essential components of the complement system have been conserved during evolution. Further hemolysins were described in the coelomic fluid of *S. nudus* (Osada et al., 1993). Naturally present antibacterial activity directed against *Vibrio alginolyticus* has been demonstrated in coelomic fluid and coelomocytes lysate of the sea urchin *Paracentrotus lividus* (Wilson et al., 1995). Only 5 min of contact was necessary for induction of substantial bactericidal activity.

Several other interesting molecules playing a role in the humoral defense of *Echinodermata* have been described. Sea star factor (SSF) is a 39 kDa protein isolated from the coelomocytes of *Asteria forbesi* and it was shown to exhibit a wide range of biological activities, such as an inflammatory reaction identical to a delayed type hypersensitivity, inhibition of macrophage migration (Prendergast and Suzuki, 1970), activation of macrophages for cytostasis (Liu et al., 1983) or suppression of T-dependent but not T-independent immune response (Prendergast et al., 1974; Willenborg and Prendergast, 1974). Subsequent studies revealed that SSF blocks the development of T-dependent antibody secreting clones by preventing lymphokine secretion (Kerlin et al., 1994). The intracytoplasmic levels of mRNA for IL-4 or IL-5 remained unaltered. The mechanisms of this inhibition are not clear and might involve either a block in effective cooperation of T and B lymphocytes or a direct effect in T lymphocytes. The facts that low doses of SSF are strongly immunosuppressive in mice without having any toxic effects and that this inhibition is fully reversible (Kerlin et al., 1994) might be potentially useful in clinical practice.

Beck and Habicht described the presence of an IL-1-like protein in the starfish *A. forbesi* (Beck and Habicht, 1986). Further characterization showed a protein of MW 22 kDa with conserved amino acid sequence when compared to mammalian IL-1 (Beck et al., 1990). This IL-1 has iso-electric points of 7.4, 5.4 and 4.8. The pI 4.8 has a MW 22 kDa, the other two have MW of 17 kDa (Beck and Habicht, 1991). Western blot analysis of the lysates using an antiserum to the human IL-1 precursor proved the presence of similar immunoreactive molecules in echinoderms. Their bio-logical activities involve stimulation of phagocytosis, activity in the human melanoma cytotoxicity assay for IL-1, stimulation of protein synthesis and PGE_2 production and growth factor activity (Beck et al., 1990; Beck and Habicht, 1991). The fact that this cytokine is so similar to vertebrate IL-1 in molecular weight, isoelectric point and biological effects suggests that IL-1 is an evolutionary stable defense molecule. The presence of IL-2 (Fig. 1) on the surface of coelomocytes was also detected (Legac et al., 1996). In addition, membrane structures similar to human receptors for IL-1, IL-2, IL-6 and IFN-γ were found in the sea star *A. rubens* (Fig. 2; Legac et al., 1996).

Non-adherent cells isolated from the axial organ of the sea star were found to react to the two subsequent challenges with the same antigen by the secretion of a specific antibody-like protein capable of lysis of hap-tenated red blood cells in the presence of complement (Delmotte et al., 1986). Compared to the immunoglobulin of the lowest vertebrates, this antibody-like substance is structurally simpler than the lamprey immuno-globulin, not having light and heavy chains.

Courtney Smith and co-workers activated sea urchin coelomocytes (*S. purpuratus*) with LPS and found that coelomocytes from sea urchin re-spond to immune challenge from LPS with significant elevation of profilin, a small actin-binding protein involved in signal transduction (Courtney Smith et al., 1995). In later experiments they tried to identify some of the expressed genes (Smith et al., 1996). They sequenced randomly chosen clones from the cDNA library to produce a set of expressed sequence tags. When the deduced amino acid sequence was compared with known protein sequences, 88 matches were found with a wide array of proteins such as an invertebrate homologue of a vertebrate complement component, clotting factors, complement receptor or regulatory proteins, C-type lectin, pro-tease inhibitors, serine protease with similarities to thrombin, elastase and plasmin, proteins involved in signaling, lysosomal proteins, and cyto-skeletal proteins (for details see Smith et al., 1996). These data represent so far the best indication of the presence of a primitive complement system composed of at least one component and a receptor in deuterostomes. This system might serve as a platform onto which the Ig gene system was later added and resulted in the sudden expansion of both systems in higher verte-brates (Smith et al., 1996).

There are other substances participating in humoral defense reactions. For instance, the red spherule cells are known to contain echinochromes (naphtaquinones) which have toxic properties on various epibiotic algae

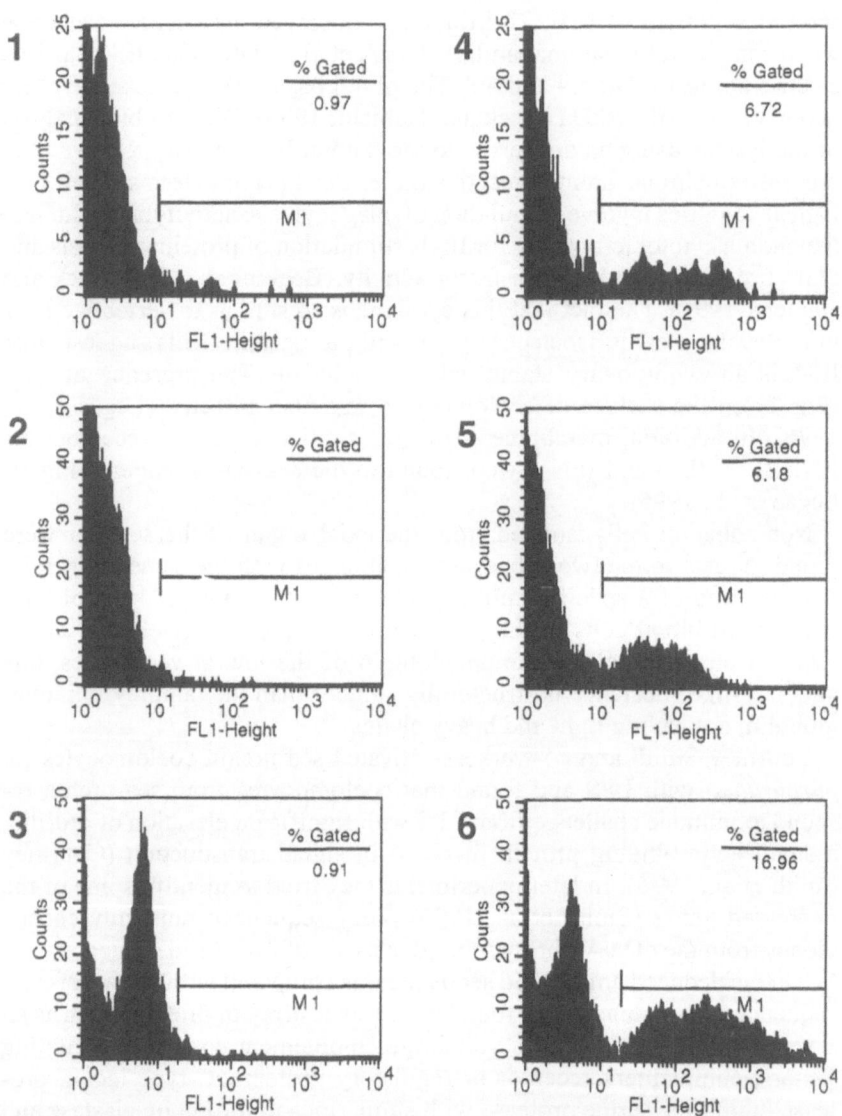

Figure 1. Cell surface detection of IL-2 on axial organ cells. Monoparametric histograms 1 to 3 of green fluorescence (murine irrelevant IgG$_1$) and histograms 4 to 6 show the results obtained with anti-IL-2 mAb. Histograms 1 and 4 illustrate data from small cells, histograms 2 and 5 from intermediate cells, and histograms 3 and 6 from large cells. Histograms 4 and 6 illustrate the presence of a small cell subset of axial organs cells with detectable cell surface IL-2. (From Legac et al., 1996, with permission.)

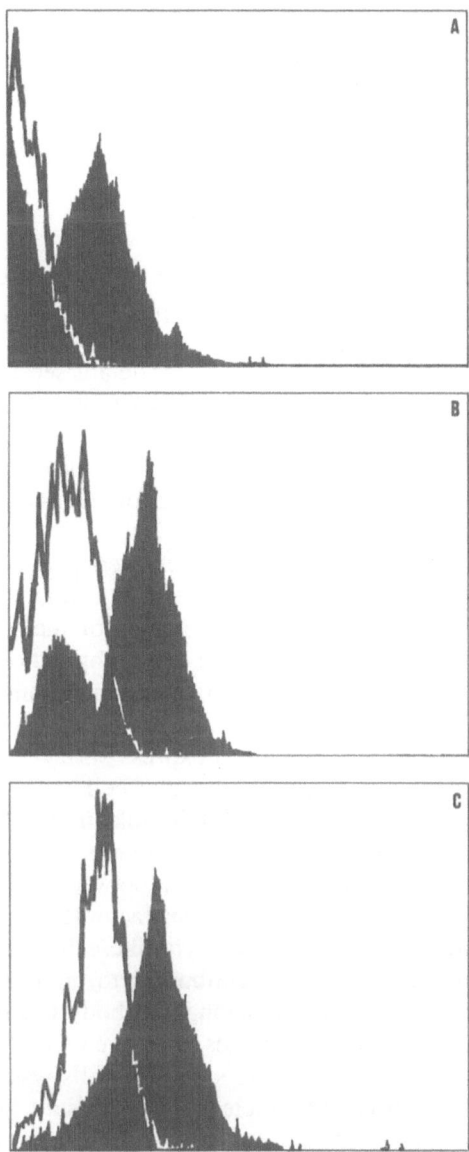

Figure 2. IL-1 receptor expression on axial organ cells. Monoparametric histogram of green fluorescence: block lines histogram represents negative control and black shadowed histogram represents results obtained with anti-IL-1 receptor mAb. Histograms A to C respectively illustrate data from small, intermediate and large cells. (From Legac et al., 1996, with permission.)

(Vevers, 1963). Melanin and its precursors, so important in immune responses of some invertebrates, have been detected in various amoebocyte types and the axial organ (Smith, 1981). Their role in echinoderm immunity is still unknown.

Cellular immunity

Six main coelomocyte categories are present in every echinoderm class. They are phagocytic amoebocytes, spherule cells, vibratile cells, hemocytes, crystal cells and progenitor cells (Boolotian and Giese, 1958; Endean, 1966; Smith, 1981; Edds, 1993; for details see Síma and Větvička, 1992). In addition, some additional cell types have been found in some representatives. As most echinoderm species were not investigated, this list is unfortunately far from being complete. The developmental origin of coelomocytes is still unknown. Candidates for coelomocytopoietic tissues are summarized in Table 2.

The axial organ of the sea star *A. rubens* is considered to be a primitive ancestral immune organ. Axial organ cells can mount both cellular and humoral responses with some characteristics of the vertebrate immune system (Leclerc et al., 1986). When the entire population was fractionated by adherence to the nylon wool, two distinct populations have been found: adherent (B-cell-like) and non-adherent (T-cell-like) (Leclerc and Bajelan, 1992). Both cell types were found to express surface molecules reacting with specific antibodies against human T-cell receptor. In additional studies, T-cell-like cells release lymphokine-like molecules with mitogenic characteristics. On the other hand, adherent cells secrete an antibody-like factor (Delmotte et al., 1986). Cooperation between lymphocyte-like cells and phagocytes is necessary to produce this factor (Leclerc et al., 1986a, b). Later studies showed that the lymphocyte-like cells can be distinguished and identified using monoclonal antibodies to human leukocytes (Leclerc et al., 1993a, b). When monoclonal antibody CD68 KP1, which recognizes various human macrophages, was used for staining of sections of the axial organ, macrophage-like subset of cells was found (Leclerc et al., 1994). Even now it is still not clear if these results are perhaps caused

Table 2. The distribution of coelomocyte forming organs among echinoderm classes

Class	Axial organ	Tiedemann's bodies	Polian vesicles	Spongy bodies
Crinoidea	+	+	−	+
Holothuroidea	−	−	+	−
Echinoidea	+	−	−	+
Asteroidea	+	+	+	−
Ophiuroidea	+	−	+	−

by a cross-reactivity between common amino acid sequences, and the possibility that sea star phagocytes share antigenic molecules with their human counterparts exists.

The main cell types playing a significant role in echinoderms defense reactions are coelomocytes and spherule cells. They are actively phago-cytizing and participating in cell clumping, clotting reaction, encapsulation, wound repair and cytotoxicity. Similar to all other animal phyla, phagocytosis is the main defense feature. Phagocytes are highly efficient (Plytycz and Seljelid, 1993) and the coelomic fluid of healthy echinoderms is almost aseptic (Bang and Lemman, 1962). The phagocytosis was highly increased by opsonization with vertebrate sera (fish, mouse, human). This opsonization required Ca^{2+} and was temperature-dependent. The involve-ment of C1 to C4 fragments of complement was suggested, as the C3-de-pleted serum was found not to be active (Bertheussen and Seljelid, 1982). A more detailed study showed that sheep red blood cells coated with human C1, C2 or C4 are not engulfed by echinoderm cells, but cells coated with human C3b were internalized rapidly (Beertheussen, 1982).

Regeneration

Echinoderms display considerable powers of regeneration. The process of regeneration is relatively long-lasting (a year or more when a larger part of the body is lost). In holothurians the eviscerated parts of digestive tract, respiratory tree (tubules of Cuvier), or tentacles are regenerated. Incisional wounds are generally healed within several days. During healing the mito-tic activity of epidermal cells seems to play a less important role than mas-sive coelomocyte accumulations. The spherule cells form a significant part of cell infiltration, producing collagen and antimicrobial substances (Men-tor and Eisen, 1973).

Clotting reaction is an especially important defense mechanism for sea urchin, as these animals cannot use muscular contraction for wound clos-ing. Degranulation or autolysis of the vibratile cells results in transforma-tion of coelomic fluid into a gel (Johnson, 1969). Alternatively, phagocytic cells form cellular clots as they form filopodia which intertwine into a net and plug the body wound (Edds, 1977, 1993).

An interesting defense feature is the formation of brown bodies. These are pigmented aggregates of amoebocytes found in the coelomic cavities of most sea cucumbers. Injection of carmine particles into the coelom of *H. tubulosa* induced formation of brown bodies. They serve as an aggregate forming an extracellular matrix upon which injected particles and amoebo-cytes collect (Jans et al., 1996). Once an aggregation is complete, destruc-tion and subsequent elimination begins.

Collaboration between spherule cells and phagocytic amoebocytes in killing of bacteria in sea urchin (*S. purpuratus* and *S. franciscanus*) has been reported (Johnson, 1969).

Spontaneous cytotoxic reaction was demonstrated in the starfish, *Pisaster gigantus* (Decker et al., 1981). These cells kill various normal cells (chicken, horse and human red blood cells) as well as tumor cells (PA313 and P815Y cells). The blocking effect of some saccharides revealed that this reaction is mediated via lectins present on the membrane of starfish cells. Leclerc and Luquet demonstrated spontaneous as well as induced cytotoxicity of axial organ cells isolated from the sea star *A. rubens* (Leclerc and Luquet, 1983; Luquet and Leclerc, 1983). The use of silica particles which selectively destroy phagocytic cells showed that the cells responsible for the spontaneous cytotoxicity against mouse myeloma cells are phagocytes (Leclerc et al., 1993). The mechanisms responsible for cytotoxic cell-mediating cell lysis in invertebrates remain unknown. In mammals, cytolytic molecules are stored in cytoplasmic granules of natural killer cells. Canicatti found a Ca-dependent, heat-labile lytic molecules in coelomic fluid isolated from sea urchin, *Paracentrotus lividus* (Canicatti, 1987). Subsequent studies showed that these molecules are stored in cytoplasmic granules of circulating coelomocytes (Pagliara and Canicatti, 1993; Pagliara et al., 1993). These hemolysins induce 10-nm lesions on target membranes (Canicatti, 1991).

In the sea cucumber *Cucumaria tricolor* and the horned sea star *Protoreaster nodosus*, the autografts are not rejected and the second-set allografts are rejected at an accelerated rate (Hildemann and Dix, 1972; Hildemann et al., 1974). Similar results have been demonstrated in the sea star *Dermasterias imbricata* (Karp and Hildemann, 1975) and in *Lytechinus pictus* (Coffaro and Hinegardner, 1977). Allograft rejection seems to operate by a nonspecific activation mechanism. The rejection kinetics of the third-party allografts are superimposable on those for the second set, even though both second-set and third-party grafts showed faster rate of rejection (Smith and Davidson, 1994). Therefore, the echinoderms cannot distinguish differences between second-set grafts to which they have been presensitized and third-party grafts that the animals have not previously been exposed to. Echinoderms (or at least sea urchins, which were used in these particular experiments) activate their immune system in response to first contact with an antigen and subsequently respond in an augmented way to any non-self antigen. The rejection of grafts is accompanied by allorecognition, which was further demonstrated *in vitro* in mixed-lymphocyte type of reaction (Bertheussen, 1979).

Conclusions

In summary, echinoderms have developed a highly efficient immune system with some interesting similarities in fundamental defense features to the chordates. The phagocytic amoebocytes are considered the analogs of the vertebrates' reticuloendothelial system, and the axial organ has similar function to the spleen or lymph nodes. Axial organ cells might be capable of nonspecific responses by means of hemolytic factors as well as highly specific inducible complement-dependent immune responses (Brillouet et al., 1986). The studies of the defense reactions of the echinodermates suggest that in this phylum the high specificity of immune mechanisms depends on some level of cooperation within the immunocompetent and accessory cells, which resembles the cooperation between lymphocytes and macrophages of vertebrates. However, the overall characteristics of their immune system are still the basis of a nonadaptive immunity of invertebrates, as they lack any adaptive immunological memory (Smith and Davidson, 1992, 1994).

References

Anderson, R.S. and Good, R.A. (1975) Naturally-occurring hemagglutinins in a tunicate, *Halocynthia pyriformis. Biol. Bull.* 148:357–369.

Bang, F.B. and Lemma, A. (1962) Bacterial infection and reaction to injury in some echinoderms. *J. Insect Pathol.* 4:401–414.

Beck, G. and Habicht, G.S. (1986) Isolation and characterization of a primitive interleukin-1-like protein from an invertebrate, *Asterias forbesi. Proc. Natl. Acad. Sci. U.S.A.* 83: 7429–7433.

Beck, G. and Habicht, G.S. (1991) Purification and biochemical characterization of an invertebrate interleukin 1. *Mol. Immunol.* 28:577–584.

Beck, G., O'Brien, R.F. and Habicht, G.S. (1990) Characterization of interleukin 1 from invertebrates. *In:* J.J. Marchalonis and C.J. Remish (eds): *Defense Molecules*, Wiley-Liss, New York, pp 125–132.

Bertheussen, K. (1979) The cytotoxic reaction in allogeneic mixtures of echinoid phagocytes. *Exp. Cell Res.* 120:373–381.

Bertheussen, K. (1982) Receptors for complement on echinoid phagocytes. II. Purified human complement mediates echinoid phagocytosis. *Dev. Comp. Immunol.* 6:635–642.

Bertheussen, K. (1983) Complement-like activity in sea urchin coelomic fluid. *Dev. Comp. Immunol.* 7:21–31.

Bertheussen, K. and Seljelid, R. (1982) Receptors for complement on echinoid phagocytes. I. The opsonic effect of vertebrate sera on echinoid phagocytosis. *Dev. Comp. Immunol.* 6: 423–431.

Boolotian, R.A. and Giese, A.C. (1958) Coelomic corpuscles of echinoderms. *Biol. Bull.* 115: 53–63.

Brillouet, C., Leclerc, M. and Luquet, G. (1986) Mitogens induced regulation of sea star axial organ cell humoral immune response *in vitro. Cell Biol.Int.Rep.* 10:667–675.

Canicatti, C. (1987) Evolution of the lytic system in echinoderms. I. Naturally-occurring hemolytic activity in *Paracentrotus lividus (Echinoidea)* coelomic fluid. *Biol. Zool.* 4:325–329.

Canicatti, C. (1991) Binding properties of *Paracentrotus lividus (Echinoidea)* hemolysin. *Comp. Biochem. Physiol.[A]* 98:463–468.

Canicatti, C. and Ciulla, D. (1987) Studies on *Holothuria polii (Echinodermata)* coelomocyte lysate. I. Hemolytic activity of coelomocyte hemolysins. *Dev. Comp. Immunol.* 11:705–712.

Canicatti, C. and Rizzo, A. (1991) A 220 kDa coelomocyte aggregating factor involved in *Holothuria polii* cellular clotting. *Eur. J. Cell Biol.* 56:79–83.

Canicatti, C., Parrinello, N. and Arizza, V. (1987) Inhibitory activity of sphingomyelin on hemolytic activity of coelomic fluid of *Holothuria polii (Echinodermata). Dev.Comp.Immunol.* 11:29–35.

Canicatti, C., Pagliara, P. and Stabili, L. (1992) Sea urchin coelomic fluid agglutinin mediates coelomocyte adhesion. *Eur. J. Cell Biol.* 58:291–295.

Clark, R.B. (1964) *Dynamics in Metazoan Evolution*. Clarendon Press, Oxford.

Coffaro, K.A. and Hinegardner, R.T. (1977) Immune response in the sea urchin *Lytechinus pictus. Science* 197:1389–1390.

Courtney Smith, L., Britten, R.J. and Davidson, E.H. (1995) Lipopolysaccharide activates the sea urchin immune system. *Dev. Comp. Immunol.* 19:217–224.

Decker, J.M., Elmholt, A. and Muchmore, A.V. (1981) Spontaneous cytotoxicity mediated by invertebrate mononuclear cells toward normal and malignant targets: inhibition by defined mono- and disaccharides. *Cell. Immunol.* 59:161–170.

Delmotte, F., Brillouet, C., Leclerc, M., Luquet, G. and Kader, J.C. (1986) Purification of an antibody-like protein from the sea star *Asterias rubens* (L.). *Eur. J. Immunol.* 16: 1325–1330.

Edds, K.T. (1977) Dynamic aspects of filopodial formation by reorganization of microfilament. *J. Cell. Biol.* 73:479–491.

Edds, K.T. (1993) Cell biology of echinoid coelomocytes. 1. Diversity and characterization of cell types. *J. Invertebr. Pathol.* 61:173–178.

Endean, R. (1966) The coelomocytes and coelomic fluid. *In:* R.A. Boolotian (ed.): *Physiology of Echinodermata*, J. Wiley and Sons, New York, pp 301–328.

Ey, P.L. and Jenkin, C.R. (1982) Molecular basis of self/non-self discrimination in the invertebrata. *In:* N. Cohen and M.M. Sigel (eds): *The Reticuloendothelial System, Vol. 3*, Plenum Press, New York, pp 321–391.

Ferguson, J.C. (1984). Translocative functions of the enigmatic organs of starfish – the axial organ, hemal vessels, Tiedemann's bodies, and rectal caeca: an autoradiographic study. *Biol. Bull.* 166:140–155.

Giga, Y. Ikai, A. and Takahashi, K. (1987) The complete aminoacid sequence of echinoidin, a lectin from the coelomic fluid of the sea urchin *Anthocidaris crasspina. J. Biol. Chem.* 262: 6197–6203.

Hall, J.D. and Rowlands, D.T. (1974) Heterogeneity of lobster agglutinins. II. Specificity of agglutinin-erythrocyte binding. *Biochemistry* 13:828–832.

Herreid, C.F., La Russa, V.F. and DeFesi, C.R. (1976) Blood vascular system of the sea cucumber, *Stichopus moebii, J. Morphol.* 150:423–451.

Hildemann, W.H. and Dix, T.G. (1972) Transplantation reaction of tropical Australian echinoderms, *Transplantation* 15:624–633.

Hildemann, W.H., Dix, T.G. and Collins, J.D. (1974) Tissue transplantation in diverse marine invertebrates. *Contemp. Top. Immunobiol.* 4:141–150.

Ito, T., Ito, Y., Osada, M. and Mori, K. (1992a) Identification of opsonin in the coelomic fluid of the sea urchin, *Strongylocentrotus nudus. Nippon Suisan Gakkaishi* 58:2119–2124.

Ito, T., Matsutani, T., Mori, K. and Nomura, T. (1992b) Phagocytosis and hydrogen peroxide production by phagocytes of the sea urchin *Strongylocentrotus nudus. Dev. Comp. Immunol.* 16:287–294.

Jans, D., Dubois, P. and Jangoux, M. (1996) Defensive mechanisms of holothuroids (*Echinodermata*): formation, role, and fate of intracoelomic brown bodies in the sea cucumber *Holothuria tubulosa. Cell Tissue Res.* 283:99–106.

Johnson, P.T. (1969) The coelomic elements of sea urchin (*Strongylocentrotus*) I. The normal coelomocytes: Their morphology and dynamics of hanging drops. *J. Invertebr. Pathol.* 13:25–41.

Karp, R.D. and Coffaro, K.A. (1982) Cellular defense system of echinodermata, *In:* N. Cohen and M.M. Sigel (eds): *The Reticuloendothelial System, Vol. 3*, Plenum Press, New York, pp 257–282.

Karp, R.D. and Hildemann, W.H. (1975) Specific rejection of integumentary allografts by the sea star *Dermasterias imbricata. Adv. Exp. Med. Biol.* 64:137–147.

Kerlin, R.L., Cebra, J.J., Weinstein, P.D. and Prendergast, R.A. (1994) Sea star factor blocks development of T-dependent antibody secreting clones by preventing lymphokine secretion. *Cell Immunol.* 156:62–76.

Leclerc, M. and Bajelan, M. (1992) Homologous antigen for T cell receptor in axial organ cells from the asterid *Asterias rubens. Cell Biol. Int. Rep.* 16:487–490.

Leclerc, M. and Luquet, G. (1983) Effect of *in vivo* inoculation of bacteria on the spontaneous cytotoxicity of axial organ cells from *Asterias rubens. Immunol. Lett.* 6:107–108.

Leclerc, M., Brillouet, C., Luquet, G. and Binaghi, R.A. (1986a) Production of an antibody-like factor in the sea star *Asterias rubens*: involvement of at least three cellular populations. *Immunology* 57:479–482.

Leclerc, M., Brillouet, C., Luquet, G. and Binaghi, R.A. (1986b) The immune system of invertebrates: the sea star *Asterias rubens (Echinoderma)* as a model of study. *Bull. Inst. Pasteur* 84:311–330.

Leclerc, M., Arneodo, V., Legac, E., Bajelan, M. and Vaugier, G. (1993a) Identification of T-like and B-like lymphocyte subsets in sea star *Asterias rubens* by monoclonal antibodies to human leukocytes. *Thymus* 21:133–139.

Leclerc, M., Bajedan, M., Barot, R. and Tlaskalova-Hogenova, H. (1993b) Effect of silica on the spontaneous cytotoxicity of axial organ cells from *Asteria rubens*. *Cell Biol. Int.* 17:787–789.

Leclerc, M., Maitre, F. and Contrepois, L. (1994) Identification of a macrophage-like subset in the axial organ sea star *Asteria rubens*. *Cell Biol. Int.* 18:835–837.

Legac, E., Vaugier, G.L., Bousquet, F., Bajelan, M. and Leclerc, M. (1996) Primitive cytokines and cytokine receptors in invertebrates: the sea star *Asteria rubens* as a model of study. *Scad. J. Immunol.* 44:375–380.

Liu, S.H., McChesney, M.B. and Prendergast, R.A. (1983) Kinetics of tumor cell cytostasis by sea star-derived macrophages. *Dev. Comp. Immunol.* 7: 545-554.

Luquet, G. and Leclerc, M. (1983) Spontaneous and induced cytotoxicity of axial organ cells from *Asteria rubens* (Asterid-Echinoderm). *Immunol. Lett.* 6:339–342.

Menton, D.N. and Eisen, A.Z. (1973) Cutaneous wound healing in sea cucumber, *Thyone briareus*. *J. Morphol.* 141:185–204.

Millott, N. (1967) *Echinoderm Biology*. Symp. Zool. Society London, Vol. 20, Academic Press, New York.

Moore, R.C. (Ed.) (1966-1978) *Treatise on Invertebrate Paleontology. Echinodermata*. Parts S – U. Geological Society of America and University of Kansas Press, Lawrence, KS.

Nichols, D. (1969) *Echinoderms*. Hutchinson Univ. Library, London.

Osada, M., Ito, T., Matsutani, T. and Mori, K. (1993) Partial purification and characterization of hemolysin from the coelomic fluid of *Strongylocentrotus nudus*. *Comp. Biochem. Physiol. [C]* 105:43–49.

Pagliara, P. and Canicatti, C. (1993) Isolation of cytolytic granules from sea urchin amoebocytes. *Eur. J. Cell Biol.* 60:179–184.

Pagliara, P., Canicatti, C. and Cooper, E.L. (1993) Structure and enzyme content of sea urchin cytolytic granules. *Comp. Biochem. Physiol. [B]* 106:813–818.

Parrinello, N., Canicatti, C. and Rindone, D. (1976) Naturally occurring hemagglutinins in the coelomic fluid of the echinoderms *Holothuria polii* Delle Chiaje and *Holothuria tubulosa* Gmelin. *Boll. Zool.* 43:259–271.

Pierschbacher, M.D. and Ruoslahti, E. (1984) Variants of the cell recognition site of fibronectin that retain attachment-promoting activity. *Proc. Natl. Acad. Sci. U.S.A.* 81:5985–5988.

Plytycz, B. and Seljelid, R. (1993) bacterial clearance by the sea urchin *Strongylocentrotus droebachiensis* – brief communication. *Dev. Comp. Immunol.* 17:283–289.

Prendergast, R.A. and Suzuki, M. (1970) Invertebrate protein stimulating mediators of delayed hypersensitivity. *Nature* 227:277–279.

Prendergast, R.A., Cole, G.A. and Henney, C.S. (1974) Marine invertebrate origin of a reactant to mammalian T cells. *Ann. NY Acad. Sci.* 234:7–16.

Šíma and Větvička, V. (1990) *Evolution of Immune Reactions*. CRC Press, Boca Raton.

Šíma and Větvička, V. (1992) Evolution of immune accessory functions. *In:* L. Fornůsek and V. Větvička (eds): *Immune System Accessory Cells*, CRC Press, Boca Raton, pp 1–55.

Smith, L.C. and Davidson, E.H. (1992) The echinoid immune system and the phylogenetic occurrence of immune mechanisms in deuterostomes. *Immunol. Today* 13:356–362.

Smith, L.C. and Davidson, E.H. (1994) The echinoderm immune system. Characters shared with vertebrate immune systems and characters arising later in deuterostome phylogeny. *In:* G. Beck, E.L. Cooper, G.S. Habicht and J.J. Marchalonis (eds): *Primordial Immunity: Foundations for the Vertebrate Immune System*, Ann. New York Acad. Science, New York, pp 213–226.

Smith, L.C., Chang, L., Britten, R.J. and Davidson, E.H. (1996) Sea urchin genes expressed in activated coelomocytes are identified by expressed sequence tags. *J. Immunol.* 156: 593–602.

Smith, V.J. (1981) The echinoderms. *In:* N.A. Ratcliffe and A.F. Rowley (eds): *Invertebrate Blood Cells, Vol. 2*, Academic Press, London, pp 513–562.

Vevers, H.G. (1963). Pigmentation of the echinoderms. *Proc. XIV Int. Congress Zool.*, 3:120–122.

Willenborg, D.O. and Prendergast, R.A. (1974) The effect of sea star coelomocyte extract on cell-mediated resistance to *Listeria monocytogenesis* in mice. *J. Exp. Med.* 139:820–833.

Wilson, M.R., van Ravenstein, E., Miller, N.W., Clem, L.W., Middleton, D.L. and Warr, G.W. (1995) cDNA sequences and organization of IgM heavy chain genes in two holostean fish. *Dev. Comp. Immunol.* 19:153–164.

Chordates/Urochordata

The members of the subphylum *Urochordata*, commonly named the tunicates, and the subphylum *Cephalochordata* are ranked together as invertebrate chordates. They form a typical marine assemblage, more remote from
the main evolutionary stream than vertebrates. They possess, however, the
three basic characteristics distinguishing chordates: the notochord, appear-
ing at least at some stage in the life cycle, the dorsal hollow nerve cord, and
the pharyngeal clefts. The members of three classes – the *Ascidiacea*, the
Thaliacea, and the *Appendicularia (Larvacea)* differ in their anatomy and
mode of life. The *Ascidiacea*, often known as the sea squirts, are the largest
grouping of urochordates to which the main interest of comparative immunologists has been devoted. In the case of the latter two classes, almost no
information about their defense mechanisms is available. Recently, a
homolog of the vertebrate brachyury (T) gene, essential for the formation
of mesoderm cells of the notochord, has been described in *Halocynthia
roretzi*, confirming the phyletic kinship of ascidians (Yasuo and Satoh,
1994). Urochordates are regarded as the evolutionary progenitors to the
vertebrate branch (Berrill, 1955, Satoh and Jeffery, 1995).

*The anatomical features of ascidians in regard to their possible immune
significance*

The coelom is not well-developed. Thus it seems that the circulatory
system, essentially a hemocoel (Berrill, 1950) with a number of various
cell types, some of them exerting highly specialized functions, has more
immunological significance in these animals. Anatomically, the vascular
system is composed of the blood vessels which are not lined by an endothelium, but represent rather the channels or lacunae running through the
connective tissue.

The immune significance of the pharyngeal region

The anterior buccal region opens internally into a large pharyngeal chamber and the walls are perforated with gill slits. The outer lining of the
pharynx is an ectodermal epithelium and the inner covering derives from
endoderm. Between these two layers the mesenchymal tissue is closed.
This structural composition of pharyngeal wall could be of further phyletic
significance. The mutual interactions among the three germinal layers
could serve as a background for later developmental potential for the
thymic emergence in vertebrates (Cooper, 1973), as does the functional
activity of endostyle (production of iodinated mucous) for the phylogeny
of thyroid gland (Frederiksson et al., 1993). The cells resident or wandering

in the pharyngeal region strongly resemble the vertebrate lymphoid cells, not only in their morphology, but also in their capacity to multiplicate after antigenic stimulation (Raftos and Cooper, 1991).

Other structures engaged in the immunity .

The lymph nodules
Even if the ascidian blood cells are renewing cell population, the ascidians have developed distinct hemopoietic structures. Ontogenetically, the stem cells arise from a mesodermal cell mass in the vicinity of archenteron (Cowden, 1968). In adults, the clusters of hemoblasts are predominantly situated in the connective mesenchymal tissues (diffused or organized into nodules), along the digestive tract (mainly within the pharyngeal wall), endostyle, and gonad. Some nodules have been identified in the body wall, too. The nodules have characteristic structure with hemoblasts in the center surrounded by cells in various differentiation stages. The final maturation of blood cells occurs in the circulation, so only few fully differentiated cells exist on the periphery of the nodule (Freeman, 1970; Ermak, 1975, 1976). Thus, the mesenchymal origin of urochordate blood cells is a common feature shared with all vertebrates (Orkin, 1995).

Body wall
The single-layered epithelium of the body surface of ascidians is covered by a tunic, a special mantle, according to which the animals obtained their common name. The unusual feature of the tunic is the presense of blood cells migrating from the deeper mesenchyme body tissues or, in those ascidians in which the tunic is supplied with blood vessels (*Ciona*), directly from the circulation (Welsch, 1984).

The neural gland
This glandular organ, anatomically located beneath the cereberal ganglion, is connected with the pharynx by way of ciliated tube. It is supposed to have exocrine functions.

Humoral immunity

Similar to other invertebrates, opsonins (often in form of lectins) play a key role in defense system of urochordates. Hemagglutinins have been studied mostly by Coombe et al. in *Botrylloides leachii* (Coombe et al., 1981, 1982, 1984). This agglutinin (called HA-2) promotes adhesion of sheep erythrocytes to mouse macrophages and is involved in particle recognition. On the other hand, naturally occurring agglutinins in *Styela plicata* and *Halocynthia hilgendorfi* do not posses any opsonic properties (Fuke and Sugai, 1972). Nevertheless, relatively little is known about other humoral molecules involved in defense mechanisms. Kelly et al. described a humoral opsonin from the solitary urochordate *S. clava* (Kelly et al., 1993a). More detailed studies revealed some functional similarities with the C-lectin. An interesting observation was the finding of a species specificity of this opsonin.

Phenoloxidase activity was demonstrated in the blood of *Ciona intestinalis*, *Ascidia mentul*, *A. virginea*, *Ascidiella scabra*, *Ascidiella aspersa*, *Polycarpa pomaria*, *Halocynthia roretzi* and *Corrella parallelogramma* (Jackson et al., 1993; Akita and Hoshi, 1995). This molecule exists inside the cells in the form of a proenzyme and for activation requires the action of proteases. It is sensitive to LPS activation and to some polysaccharides. As the phenoloxidase activity was detected in all ascidian species tested, it appears that it is widely distributed in tunicates. Recent studies demonstrated that hemocytes release phenoloxidase upon contact reaction. The contact unleashes so-called exocytotic burst which is, unlike in mammalian phagocytes, Ca^{2+} dependent and not connected to production of H_2O_2 or superoxide (Akita and Hoshi, 1995).

Various stimulants induced the release of an interesting metallo-protease from hemocytes of *H. roretzi*. The stimulants include LPS, calcium ionophore, thrombin, concanavalin A and phorbol myristate acetate, but not β-glucan (Coombe et al., 1981). Similar LPS-sensitive protease was found in the blood cells of *C. intestinalis* (Jackson and Smith, 1993). Its activity was induced by LPS, laminarin and some carbohydrates. A lipopolysaccharide-binding hemagglutinin isolated from cells of *H. roretzi* with a MW of 120 kDa binds not only to horse erythrocytes, but also to various Gram-negative as well as Gram-positive bacteria (Azumi et al., 1991). It also recognizes bacterial lipopolysaccharide. The inhibition studies showed that this molecule is a protein different from the plasma lectins on the basis of molecular weight and sugar specificity. The ability to agglutinate marine bacteria points toward its possible role in defense against bacterial invasion.

A putative homolog of human EB1, the protein which binds adenomatous polyposis coli, has been cloned from *Botryllus schlosseri* (Pancer et al., 1996). This protein shares 48% of the residues in human EB1. The authors hypothesized that such significant degree of conservation throughout the evolution suggests an essential regulatory mechanism.

Cloning of a urochordate cDNA revealed a 50% similarity with the short consensus repeats of the human complement control superfamily, most of all factor H, apolipoprotein H, selectins and complement receptor type 1 and 2 (Pancer et al., 1995). This mosaic and polymorph botryllid sequence might be an ancestral molecule in the evolution of the chordate's complement control protein superfamily (Reid et al., 1986; Reid and Day, 1989).

Several molecules with cancerostatic properties have been isolated from ascidians (Fukuzawa et al., 1994; for review see Watters and Vandenbrenk 1993). Esteinascidin 729 is one such pharmacologically active esteinascidin (Guan et al., 1993) and is isolated from the marine tunicate *Ecteinascidia turbinata*. It has significant antitumor effects and is already under evaluation by the National Cancer Institute (Reid et al., 1996). *In vivo* studies showed the inhibition of leukemia and melanoma cells implanted in mice.

Another interesting molecule isolated from tunicates are polyaromatic alkaloids called lamellarins found in the genus *Didemnum*. Cytotoxicity experiments of 13 different lamellarins showed their toxicity towards mouse and human tumor cell lines (Quesada et al., 1996). Further, lipophilic cyclic peptides isolated from ascidians have been receiving pharmaceutical interest because of the high potency of their cytotoxic and/or antineoplastic activities. For instance, several cytotoxic cyclic peptides containing unusual thiazole and oxazoline rings have also been isolated: the ascidiacyclamide, the ulithiacyclamide, and several types of patellamides. All these cyclic peptides have similar structure responsible for their cytotoxic activity (Ishida et al., 1992; In et al., 1993). Moderate cytotoxicity and antimicrobial activity are exhibited also by some of the crucigasterins (polysaunsaturated amino alcohols), isolated from the *Pseudodistoma crucigaster* (Jares-Erijman et al., 1993). Potent antimicrobial properties against Gram-positive bacteria have been found for piclavines from *C. picta* (Raub and Cardelina, 1992). Other new bioactive substances, such as alkaloids, macrolides, polyether, terpenoids, carotenoids and miscellanous toxins are reviewed by Watters and van den Brenk (1993). More importantly, some of these compounds displayed inhibition of multidrug resistance or increased the toxicity of cancerostatic drugs in multidrug resistant lines. As multidrug resistance becomes an increasingly a serious problem in human medicine, further characterization of lamellarins clearly deserves our attention.

Cellular immunity

Tunicate blood contains several morphologically distinct cell types (Freeman, 1970; Milanesi and Burighel, 1978; Ermark, 1982). The classification of ascidian hemocytes remains unclear, despite some attempts (Wright, 1981), and confusion still exists both in terminology and functions. For detailed classification and characterization of hemocytes in *S. clava*, see Sawada et al. (1993). *In vitro* cultures were not too successful (Warr et al., 1977; Raftos et al., 1987a). More promising results were achieved by Rinkevich and Rabinowitz (1993). They cultivated blood cells from *B. schlosseri* for up to 3 months. It is clear that further research of ascidian cell type would greatly benefit from the availability of established cell lines or *in vitro* cultures of blood cells.

Recently, De Leo has proposed new terminology of ascidian cell types from the point of view of their morphology and immune functions (1992).

The most important cell type is considered to be the lymphocyte, known also by names such as hemoblast, hemocytoblast, lymphocyte-like cells, stem cell or progenitor cell. The participation in encapsulation, allograft rejection, cytotoxicity, proliferative response and immune memory is attributed to these cells.

Phagocytosis remains the fundamental line of defense (Dan-Sohkawa et al., 1995). The numerous lectins potentially serving as opsonins have been described in tunicates, but the direct proof of their role in antigen recognition is still lacking (Coombe et al., 1981, Fuke and Sugai, 1972). When phagocytosis by hemocytes isolated from *B. schlosseri* was studied in full detail (Fig. 3), it was found that the reaction is influenced by temperature, pH, concentration and physicochemical properties of the prey (Ballarin et al., 1994). Respiratory burst, indicated by increased production of superoxide, was associated with phagocytosis. All these results suggest close similarities between phagocytic processes of vertebrates and their closely related invertebrate relatives (Field et al., 1988). Surprisingly, the phagocytosis occurred independently of any humoral opsonizing factors. Other studies showed the presence of opsonic factors in ascidian hemolymph. These factors act by recognizing and binding target cells, as the absorption with target cells blocked the opsonic activity (Raftos, 1994). The recognition of targets by opsonic factors can be inhibited by some polysaccharides and EDTA (Kelly et al., 1993b).

Takahashi et al. prepared monoclonal antibodies that inhibit cellular reaction between hemocytes in *H. roretzi*. One of the antibodies, named A74, also inhibited phagocytosis of foreign material. Immunocytochemistry observations revealed the localization of the A74 antigen on hemocytes. After isolation and purification, the A74 protein was characterized as a 160 kDa glycoprotein with no similarity to other known proteins (Takahashi et al., 1995). This protein might be potentially important for elucidation of recognition mechanisms in tunicates.

The exceptional property of ascidians, not found in any other chordates, is their capacity of blastogenesis, a type of an asexual reproduction, which can be connected to the remarkable regeneration power of these animals. In the renewal processes, the blood cells play a significant role (Freeman, 1970). During the budding, in the central vesicles derived from the mesenchymal septum of the central vessels of the stolon, the lymphocyte-like cells accumulate forming aggregations from which the differentiation of new zooids begins (Wright, 1981). The collaboration of all three embryonic layers with blood cells in regeneration processes has been described in *Botrylloides* by Rinkevich et al. (1995). In these creatures, every minute fragment of peripheral blood vessel containing blood cells has potential to give rise to a fully developed animal.

In the solitary tunicate *S. plicata*, autografts survive indefinitely, while allografts are rejected (Raftos, 1994). The rejection is started by recognition of allogeneic cellular determinants followed by the cytotoxic destruc-

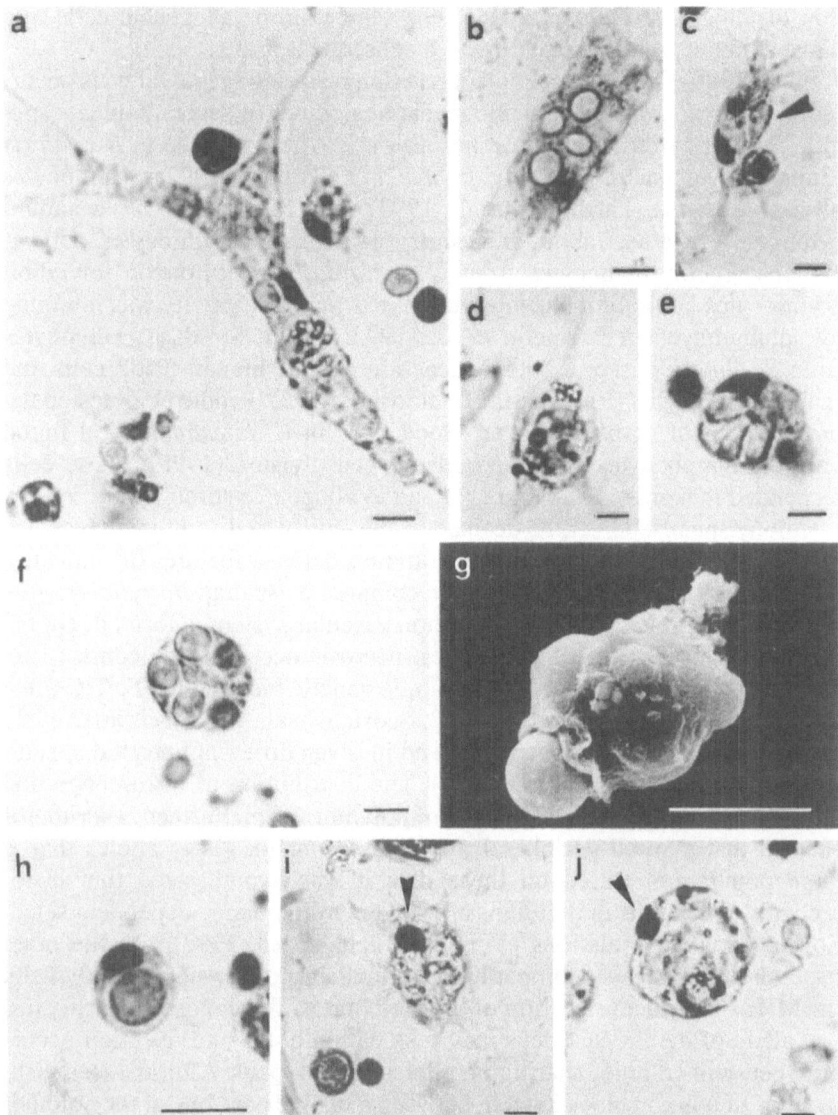

Figure 3. Hemocytes involved in phagocytosis. a–d: hyaline amoebocytes containing yeast cells (a, b), SRBC (c, arrowhead), and latex beads (d); e–f: macrophage-like cells filled with SRBC (e) and yeast cells (f); g: scanning electron micrograph of a hemocyte engaged in ingesting a yeast cell: lumps on the blood cell surface are already ingested yeast cells; h: signet-ring cell with a yeast particle inside its vacuole; i–j: yeast-containing hemocytes, showing dark formazan deposits after NBT treatment (arrowheads). Bar length: 5 μm. (From Ballarin et al., 1994, with permission.)

tion of allogeneic cells (Raftos, 1990). Injection of allogeneic cells into naïve animals preimmunized them to subsequent grafts.

Fuke (1980) found that a contact reaction resulting in mutual lysis occurred between xenogeneic and allogeneic hemocytes of several tunicate species. Similar cytotoxic reaction has also been demonstrated in *B. scalaris* (Saito and Watanabe, 1982) and in *Styela* (Kelly et al., 1993a). Hemocytes of *C. intestinalis* (Parrinello et al., 1993) were found to posses a natural cytotoxicity against rabbit, pig, sheep and human erythrocytes. Univacuolar refractile hemocytes from *C. intestinalis* are cytotoxic for rabbit erythrocytes but not for sheep erythrocytes, and this activity was inhibited by sphingomyelin (Parrinello et al., 1995, 1996). Similarly, hemocytes from *S. plicata* posses cytotoxic reaction against human K562 cells and rabbit erythrocytes (Raftos and Hutchinson, 1995). Peddie et al. tested the proliferation of undifferentiated blood cells in *C. intestinalis* and found only the lymphocyte-like cells undergo cell division (1995). These cells responded to Con A, PHA and LPS and to allogeneic stimulation.

Colony specificity can be defined as the ability to discriminate between self and non-self. It is one of the common defense features of tunicates. This reaction has been studied in the compound ascidian *Botrylloides fuscus*. In compatible colonies, a common vascular system is formed. On the other hand, in incompatible colonies, necrosis occurs at the contact site. The colony specificity resides in a single genetic locus called Fu/HC (Oka and Watanabe, 1960; Sabbadin, 1962; Scofield et al., 1982, Weissman et al., 1990). This reaction has been observed in seven different botryllid species (for review see Hirose et al., 1994). The distribution of historecognition alleles has an extensively high level of polymorphism. Further experiments showed that a small peripheral population contains fewer alleles than a large population. Based on these data it was hypothesized that historecognition systems in ascidians are subject to frequency dependent selection in natural populations (Yund and Feldgarden, 1992). On the other hand, the rules of histocompatibility in tunicates differ significantly from the MHC-dependent rejection of the vertebrates. Two colonies sharing just one allele of the Fu/HC locus may fuse within hours and rejection occurs only between colonies sharing no alleles of that locus. After the establishment of a common blood system between a fusible pair, one of the colonies in the newly formed chimera is usually resorbed. Rinkevich and Weissman found that allogeneic resorption is not only genetically controlled, but is also controlled by a polymorphic hierarchial (1993) phenomenon including the Fu/HC locus and additional unrelated loci (Rinkevich and Weissman, 1992a, b; Rinkevich, 1993; Rinkevich et al., 1994a, b). Subsequent studies revealed no significant allospecific memory (Rinkevich and Weissman, 1990; Rinkevich et al., 1994b). (For detailed description of allorecognition and xenorecognition in *Botrylloides*, see Rinkevich and Weissman, 1991; Rinkevich and Saito, 1992; Rinkevich et al., 1992, 1994a; for review see Rinkevich, 1992.)

Hirose et al. observed in *Botrylloides fuscus* that after artificial cuts, the surface contacts between incompatible colonies resulted in an interesting formation of chimeras with common vascular system (Fig. 4), called surgical fusion (1994). This observation is in contrast to the rejection occurring under the same conditions in other botryllids. The authors speculate that self/non-self recognition in tunicates is the specialized feature of tunic cells and their precursors. The fusion might be a different process than rejection with quite different recognition mechanisms.

Nonfusion reactions and rejection responses between allogeneic tunicates are associated with heavy infiltration by blood cells and subsequent cytotoxic activity (Raftos et al., 1987a; Rinkevich, 1992). Similar results were found in the solitary urochordate *S. clava* (Kelly et al., 1992). Cellular contacts were necessary for the killing and the reaction was not sensitive to temperature changes. This reaction showed surprisingly high level of response at low allogeneic ratios and the percentage cytotoxicities are consistent with multiple killing by discrete immunocompetent cell populations. The lymphocyte-like cells have been suggested to be the effector cells (Raftos et al., 1987b).

When this cytotoxic reaction was studied *in vitro* against mammalian target cells, hemocytes of *C. intestinalis* were found to possess significant cytotoxic activity against mouse (WEHI, P815 and L929) and human (K562) cell lines (Peddie and Smith, 1994). Comparison of different hemocyte populations revealed that only phagocytic and nonphagocytic amoebocytes are cytotoxic (Peddie and Smith, 1993).

S. clava 17.5 kDa opsonin exhibits some cytokine-like properties in tunicates. It acts as a powerful chemoattractant and increases incorporation of ^{3}H-thymidine into cultures of pharyngeal tissues (Kelly et al., 1993b). The production is stimulated by zymosan particles and activates tunicate phagocytes to ingest latex beads. Based on these properties, it was considered to be identical with previously described tunicate IL-1β molecule (Beck et al., 1989; Raftos et al., 1991), which was found to stimulate proliferation of mouse T lymphocytes. Besides IL-1β molecule, another opsonic protein designated IL-1α has been described (Beck et al., 1989). Both molecules significantly enhanced phagocytosis of yeast or latex beads by amoebocytes (Beck et al., 1993). The experimental data suggests that these two molecules are of similar structure, whereas the IL-1β molecule is the predominant form. IL-1β, but not IL-1α, stimulates the proliferation of tunicate cells. On the other hand, both cytokines stimulate the cell migration (Kelly et al., 1993 a). IL-1 was found in all eight species of tunicate tested (Beck et al., 1989). The number and extent of all physiological roles of IL-1 in ascidians is still not clear. IL-1β is apparently constitutively present in the hemolymph, thus the contribution to hematopoiesis was suggested (Raftos et al., 1991). Based on these results, IL-1 has been proposed as a central regulator of defense reaction in tunicates (Raftos, 1994). The fact that IL-1 was also isolated from echinoderms (Beck and Habicht,

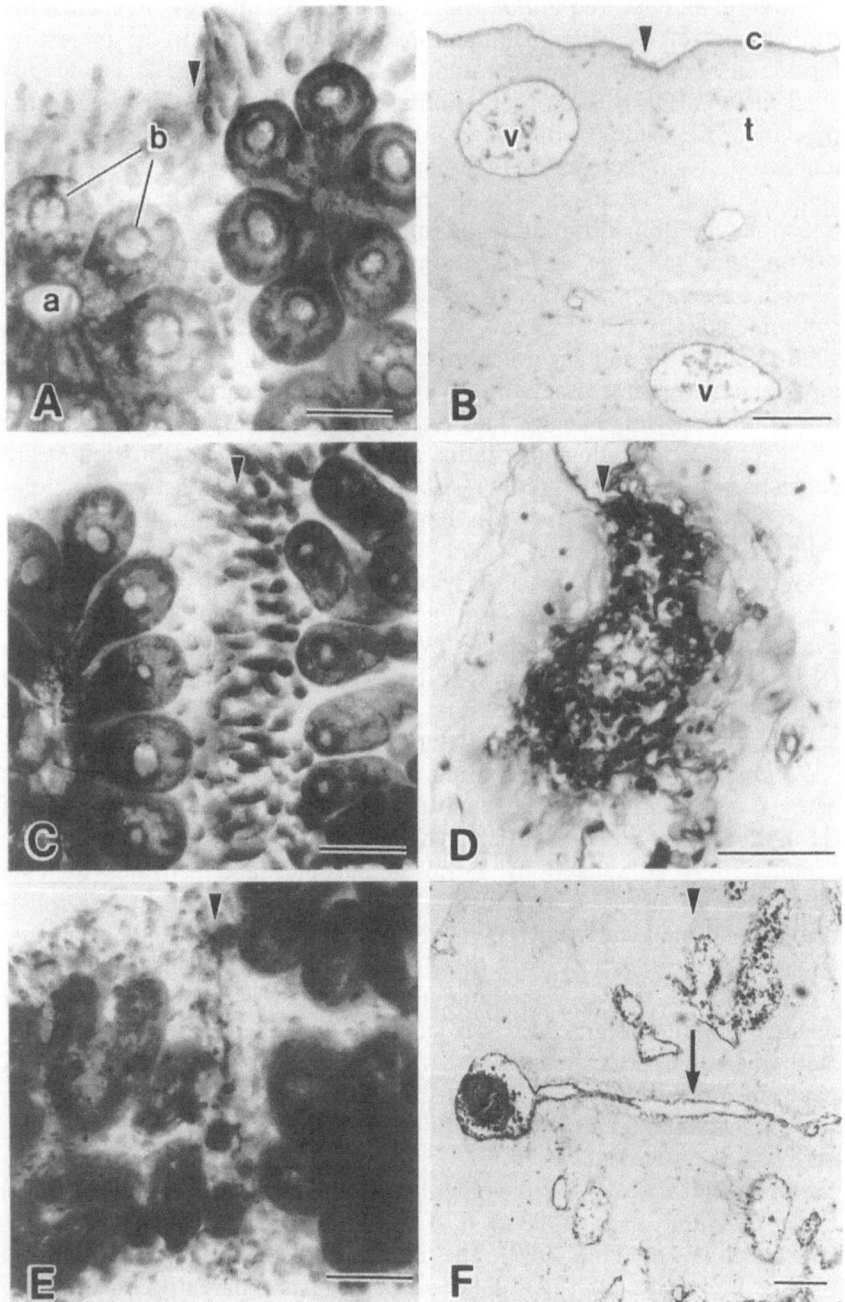

Figure 4. Occurrence of colony specificity in *Botrylloides fuscus*. Reaction between colonies, 2 days after contact. Intact colonies (A, C, E) and histological sections (B, D, F). Arrowheads indicate the contact planes. a – atrial aperture; b – branchial siphon; c – cuticle of tunic; t – tunic matrix; v – blood vessel.

1986) suggests that IL-1 is an evolutionary conserved defense molecule and points to its importance in the evolutionary development of host defense mechanisms. Another interesting observation was made when tunicate hemocytes were tested for the presence of Lyt antigens – they were found to express an Lyt-2/3 homolog of murine Lyt-2/3 complex (Negm et al., 1992).

The possibility to culture tunicate pharyngeal cells (Raftos et al., 1990) allowed the studies of the effect of human cytokines and mitogens on tunicate cells. Human IL-2 stimulated proliferation of tunicate cells in a dose-dependent manner and this stimulation was inhibited by anti-IL-2 antibodies (Raftos et al., 1991). Of three mitogens tested (PHA, Con A and PWM), only PHA was found to be mitogenic. Con A inhibited hemocyte aggregation, while WGA and ricin inhibited the allogeneic reaction. Phagocytosis was stimulated by ricin, but unchanged by Con A or WGA (Azumi et al., 1996). These data suggest that different carbohydrates are involved in regulation of the cellular reaction and that the activities of at least some cytokines have been conserved during phylogenesis.

Lymphocyte-like cells of *S. clava* proliferate in response to allogeneic stimuli. This proliferation was restricted to discrete crypts of dividing cells within the body wall of recipients (Raftos et al., 1991). These data are contrary to the previously suggested lack of immunological memory in ascidians. Raftos and Cooper hypothesize that adaptive histoincompatibility response in tunicates depends on the specific proliferation of immuno-competent cells (Raftos and Cooper, 1991).

Figure 4. (continued)
Fusion at the growing edge (A, B). Colonies have become a single mass (A). The boundary between colonies has completely disappeared and has been filled with tunic matrix (B). The cells distributed in the tunic matrix are tunic cells. Rejection at the growing edge (C, D). Ampullae of both colonies are pushing against each other (C). Necrosis is visible in the tunic at the contact boundary (D). Surgical fusion at the cut surface (E, F). Colonies having a common vascular system were photographed through the substratum (E). The histological characteristics of surgical fusion (F) are not different from those of fusion at the growing edge between compatible colonies (matching fusion). Arrow in B indicates a blood vessel crossing the contact plate.
Scale bars: 0.5 mm for A, C, E; 100 μm for B, F; 50 μm for D. (From Hirose et al., 1994, with permission.)

Conclusions

Tunicates, as the highest evolved contemporary deuterostome invertebrates, occupy a key position on the evolutionary line toward vertebrates.
They are phylogenetically primitive chordates, thus, the probability that
they share defenses characteristics of both invertebrates and chordates is
significant. Although modern tunicates are probably quite far from the
main evolutionary stem of the chordates, this group is still the closest group
of vertebrates.

References

Akita, N. and Hoshi, M. (1995) Hemocytes release phenoloxidase upon contact reaction, an allo-
    geneic interaction, in the Ascidian *Halocynthia roretzi*. *Cell Structure Function* 20: 81–87.
Azumi, K., Ozeki, S., Yokosawa, H. and Ishii, S. (1991) A novel lipopolysaccharide-binding
    hemagglutinin isolated from hemocytes of the solitary ascidian, *Halocynthia roretzi*: It can
    agglutinate bacteria. *Dev. Comp. Immunol.* 15:9–16.
Azumi, K., Takahashi, H., Ishimoto, R. and Yokosawa, H. (1996) Effects of plant lectins on cel-
    lular defense reactions of ascidian hemocytes. *Experientia* 52:839–842.
Ballarin, L., Cima, F. and Sabbadin, A. (1994) Phagocytosis in the colonial ascidian *Botryllus
    schlosseri*. *Dev. Comp. Immunol.* 18:467–481.
Beck, G. and Habicht, G.S. (1986) Isolation and characterization of a primitive interleukin-1-
    like protein from an invertebrate, *Asterias forbesi*. *Proc. Natl. Acad. Sci. U.S.A.* 83:
    7429–7433.
Beck, G., Vasta, G.R., Marchalonis, J.J. and Habicht, G.S. (1989) Characterization of inter-
    leukin-1 activity in tunicates. *Comp. Biochem. Physiol. [B]* 92:93–98.
Beck, G., O'Brien, R.F., Habicht, G.S., Stillman, D.L., Cooper, E.L. and Raftos, D.A. (1993)
    Invertebrate cytokines III: Invertebrate interleukin 1-like molecules stimulate phagocytosis
    by tunicate and echinoderm cells. *Cell. Immunol.* 146:284–299.
Berrill, N.J. (1950) *The Tunicata with an Acount of the British Species*. Ray Society, London.
Berrill, N.J. (1955) *The Origin of Vertebrates*. Oxford University Press, New York.
Coombe, D.R., Ey, P.L., Schluter, S.F. and Jenkin, C.R. (1981) An agglutinin in the haemolymph
    of an ascidian promoting adhesion of sheep erythrocytes to mouse macrophages. *Immunolo-
    gy* 42:661–669.
Coombe, D.R., Schluter, S.F., Ey, P.L. and Jenkin, C.R. (1982) Identification of the HA-2 agglu-
    tinin in the haemolymph of the ascidian *Botrylloides leachii* as the factor promoting adhesi-
    on of sheep erythrocytes to mouse macrophages. *Dev. Comp. Immunol.* 6:65–74.
Coombe, D.R., Ey, P.L. and Jenkin, C.R. (1984) Particle recognition hy haemocytes from the
    colonial ascidian *Botrylloides leachii*: evidence that the *B. leachii* HA-2 agglutinin is opso-
    nic. *J. Comp. Physiol. Bull.* 154:509–521.
Cooper, E.L. (1973) The thymus and lymphomyeloid system in poikilothermic vertebrates. *In:*
    A.J.S. Davies and R.L. Carter (eds): *Contemporary Topics in Immunobiology 2*, Plenum
    Press, New York, pp 13–38.
Dan-Sohkawa, M., Morimoto, M. and Kaneko, H. (1995) *In vitro* reactions of coelomocytes
    against sheep red blood cells in the solitary ascidian *Halocynthia roretzi*. *Zool. Sci.* 12:
    411–417.
Ermark, T.H. (1975) An autoradiographic demonstration of blood cell renewal in *Styela clava
    (Urochordata: Ascidiacea)*. *Experientia* 31:837–839.
Ermark, T.H. (1976) The hematogenic tissues of tunicates. *In:* R.K. Wright and E.L. Cooper
    (eds): *Phylogeny of Thymus and Bone Marrow-bursa Cells*, Elsevier/North Holland, Amster-
    dam, pp 45–56.
Ermark, T.H. (1982) The renewing cell population of ascidians. *Acta Zool.* 22:795–805.
Field, K.G., Olsen, G.J., Lane, D.J., Giovannoni, S.J., Ghiselin, M.T., Raff, E.C., Pace, N.R. and
    Raff, R.A. (1988) Molecular phylogeny of the animal kingdom. *Science* 239:748–753.

Freeman, G. (1970) The reticuloendothelial system of tunicates. *J. Reticuloendothel. Soc.* 7: 183–194.

Frederiksson, G., Lebel, J.M. and Leloup, J. (1993) Thyroid hormones and putative nuclear T3 receptor in tissues of the ascidian, *Phalussia mammillata* Cuvier, *Gen. Comp. Endocrinol.* 92:379–387.

Fuke, M.T. (1980) "Contact reaction" between xenogeneic or allogeneic coelomic cells of solitary ascidians. *Biol. Bull.* 158:304–315.

Fuke, M.T. and Sugai, T. (1972) Studies on the naturally occurring hemagglutinin in the coelomic fluid of an ascidian. *Biol. Bull.* 143:140–149.

Fukuzawa, S., Matsunaga, S. and Fusetani, N. (1994) Ritterazine A, a highly cytotoxic dimeric steroidal alkaloid from the tunicate *Ritterella tokioka*. *J. Org. Chem.* 59:6164–6166.

Guan, Y., Sakai, R., Rinehart, K.L. and Wang, A.H.J. (1993) Molecular and crystal structures of ecteinascidins – potent antitumor compounds from the caribean tunicate *Esteinascidia turbinata*. *J. Biomol. Struct. Dynam.* 10:793–818.

Hirose, E., Saito, Y. and Watanabe, H. (1994) Surgical fusion between incompatible colonies of the compound ascidian, *Botrylloides fuscus*. *Dev. Com. Immunol.* 18:287–294.

In, Y., Doi, M., Inoue, M., Ishida, T., Hamada, Y. and Shioiri, T. (1993). Molecular conformation of patellamide A, a cytotoxic cyclic peptide from the ascidian *Lissoclinum patella*, by x-ray crystal analysis. *Chem. Pharm. Bull.* 4:1686–1690.

Ishida, T., In, Y., Doi, M., Inoue, M., Hamada, Y. and Shioiri, T. (1992). Molecular conformation of ascidiacyclamide, a cytotoxic cyclic peptide from ascidian: X-ray analyses of its free and solvate crystals, *Biopolymers* 13:130–143.

Jackson, A.D. and Smith, V.J. (1993) LPS-sensitive protease activity in the blood cells of the solitary ascidian, *Ciona intestinalis* (L). *Comp. Biochem. Physiol. [B]* 106:505–512.

Jackson, A.D., Smith, V.J. and Peddie, C.M. (1993) *In vitro* phenoloxidase activity in the blood of *Ciona intestinalis* and other ascidians. *Dev. Comp. Immunol.* 17:97–108.

Jares-Erijman, E.A., Bapat, C.P., Ligthgow-Bertelloni, A., Rinehart, K.L. and Sakai, R. (1993). Crucigasterins, new polyunsaturated amino alcohols from the Mediterranean Tunicate *Pseudodistoma crucigaster*. *J. Org. Chem.* 58:5732–5737.

Kelly, K.L., Cooper, E.L. and Raftos, D.A. (1992) *In vitro* allogeneic cytotoxicity in the solitary urochordate *Styela clava*. *J. Exp. Zool.* 262:202–298.

Kelly, K.L., Cooper, E.L. and Raftos, D.A. (1993a) A humoral opsonin from the solitary urochordate *Styela clava*. *Dev. Comp. Immunol.* 17:29–39.

Kelly, K.L., Cooper, E.L. and Raftos, D.A. (1993b) Cytokine-like activities of a humoral opsonin from the solitary urochordate *Styela clava*. *Zool. Sci.* 10:57–64.

Milanesi, C. and Burighel, P. (1978) Blood cell ultrastructure of the ascidian *Botryllus schlosseri*. I. hemoblasts, granulocytes, morula cells and nephrocytes. *Acta Zool.* 132: 229–243.

Negm, H.I., Mansour, M.H. and Cooper, E.L. (1992) Identification and structural characterization of an Lyt-2/3 homolog in tunicates. *Comp. Biochem. Physiol.[B]* 101:55–67.

Oka, H. and Watanabe, H. (1960) Problems of colony-specificity in compound ascidians. *Bull. Marine Biol. Station Asamushi* 10:153–155.

Pancer, Z., Gershon, H. and Rinkevich, B. (1995) Cloning of a urochordate cDNA featuring mammalian short consenses repeats (SCR) of complement-control protein superfamily. *Comp. Biochem. Physiol.[B]* 111:625–632.

Pancer, Z., Cooper, E.L. and Muller, W.E.G. (1996) A urochordate putative homolog of human EB1, the protein which binds APC. *Cancer Lett.* 109:155–160.

Parrinello, N., Arizza, V., Cammarata, M. and Parrinello, D.M. (1993) Cytotoxic activity of *Ciona intestinalis (Tunicata)* hemocytes: properties of the *in vitro* reaction against erythrocyte targets. *Dev. Comp. Immunol.* 17:19–27.

Parrinello, N., Cammarata, M., Lipari, L. and Arizza, V. (1995) Sphingomyelin inhibition of *Ciona intestinalis* (Tunicata) cytotoxic henocytes assayed against sheep erythrocytes. *Dev. Comp. Immunol.* 19:31–41.

Parrinello, N., Cammarata, M. and Arizza, V. (1996) Univacuolar refractile hemocytes from the tunicate *Ciona intestinalis* are cytotoxic for mammalian erythrocytes in vitro. *Biol. Bull.* 190: 418–425.

Peddie, C.M. and Smith, V.J. (1993) *In vitro* spontaneous cytotoxic activity against mammalian target cells by the hemocytes of the solitary ascidian, *Ciona intestinalis*. *J. Exp. Zool.* 267:616–623.

Peddie, C.M. and Smith, V.J. (1994) Blood cell-mediated cytotoxic activity in the solitary ascidian *Ciona intestinalis*. *In:* G. Beck, E.L. Cooper, G.S. Habicht and J.J. Marchalonis (eds):

Primordial Immunity: Foundations for the Vertebrate System, Ann. New York Acad. Science, New York, pp 332–370.

Peddie, C.M., Riches, A.C. and Smith, V.J. (1995) Proliferation of undifferentiated blood cells from the solitary ascidian, *Ciona intestinalis, in vitro. Dev. Comp. Immunol.* 19:377–387.

Quesada, A.R., Gravalos, G. and Fernandez Puentes, J.L. (1996) Polyaromatic alkaloids from marine invertebrates as cytotoxic compounds and inhibitors of multidrug resistance caused by P-glycoprotein. *Br. J. Cancer* 74:677–682.

Raftos, D.A. (1990) Cellular restriction of histocompatibility responses in the solitary urochordate *Styela plicata* (Urochordata, Ascidiacea). *Cell Immunol.* 15:241–249.

Raftos, D.A. (1994) Allorecognition and humoral immunity in tunicates. *In:* G. Beck, E.L. Cooper, G.S. Habicht and J.J. Marchalonis (eds): *Primordial Immunity: Foundations for the Vertebrate System,* Ann. New York Acad. Science, New York, pp 227–244.

Raftos, D.A. and Cooper, E.L. (1991) Proliferation of lymphocyte-like cells from the solitary tunicate, *Styela clava*, in response to allogeneic stimuli. *J. Exp. Zool.* 260:391–400.

Raftos, D.A. and Hutchinson, C.A. (1995) Cytotoxicity reactions in the solitary tunicate *Styela plicata. Dev. Comp. Immunol.* 19:463–471.

Raftos, D.A., Tait, N.N. and Brisco, D.A. (1987a) Cellular basis of allograft rejection in the solitary urochordate *Styela plicata. Dev. Comp. Immunol.* 11:713–725.

Raftos, D.A., Tait, N.N. and Brisco, D.A. (1987b) Allograft rejection and alloimmune memory in the solitary urochordate, *Styela plicata. Dev. Comp. Immunol.* 11:343–351.

Raftos, D.A., Stillman, D.L. and Cooper, E.L. (1990) *In vitro* culture of tissue from the tunicate, *Styela clava. In Vitro* 26:962–970.

Raftos, D.A., Cooper, E.L., Habicht, G.S. and Beck, G. (1991) Invertebrate cytokines: Tunicate cell proliferation stimulated by an interleukin 1-like molecule. *Proc. Natl. Acad. Sci. U.S.A.* 88:9518–9522.

Raub, M.F. and Cardellina, J.H. (1992). The piclavines, antimicrobial indolizidines from the tunicate *Clavlina picta. Tetrahedron Lett.* 33:2257–2260.

Reid, K.B.M., Bentley, D.R., Campbell, R.D., Chung, L.P., Sim, R.B., Kristensen, T. and Tack, B.F. (1986) Complement system proteins which interact with C3b or C4b. *Immunol. Today* 7:230–234.

Reid, K.B.M. and Day, A.J. (1989) Structure-function relationships of the complement components. *Immunol. Today* 10:177–180.

Reid, J.M., Walker, D.L. and Ames, M.M. (1996) Preclinical pharmacology of ecteinascidin 729, a marine natural product with potent antitumor activity. *Cancer Chemother. Pharmacol.* 38:329–334.

Rinkevich, B. (1992) Aspects of the incompatibility nature in botryllid ascidians. *Anim. Biol.* 1:17–28.

Rinkevich, B. (1993) Immunological resorption in *Botryllus schlosseri* (Tunicata) chimeras is characterized by multilevel hierarchial organization of histocompatibility alleles. A speculative endeavor. *Biol. Bull.* 184:342–345.

Rinkevich, B. and Rabinowitz, C. (1993) *In vitro* culture of blood cells from the colonial protochordate *Botryllus schlosseri. In Vitro Cell. Dev. Biol.* 29A:79–85.

Rinkevich, B. and Saito, Y. (1992) Self-nonself recognition in the colonial protochordate *Botryllus schlosseri* from Mutsu Bay, Japan. *Zool. Sci.* 9:983–988.

Rinkevich, B. and Weissman, I.L. (1990) Failure to find alloimmune memory in the resorption phenomenon of *Botryllus cytomictical* chimera. *Eur. J. Immunol.* 20:1775–1779.

Rinkevich, B. and Weissman, I.L. (1991) Interpopulational allogeneic reaction in the colonial protochordate *Botryllus schlosseri. Int. Immunol.* 3:1265–1272.

Rinkevich, B. and Weissman, I.L. (1992a) Allogeneic resorption in colonial protochordate. Consequences of nonself recognition. *Dev. Comp. Immunol.* 16:275–286.

Rinkevich, B. and Weissman, I.L. (1992b) Incidents of rejection and indifference in Fu/HC incompatible protochordate colonies. *J. Exp. Zool.* 263:105–111.

Rinkevich, B., Shapira, M., Weissman, I.L. and Saito, Y. (1992) Allogeneic responses between three remote populations of the cosmopolitan ascidian *Botryllus schlosseri. Zool. Sci.* 9:989–994.

Rinkevich, B., Saito, Y. and Weissman, I.L. (1993) A colonial invertebrate species that displays a hierarchy of allorecognition responses. *Biol. Bull.* 184:79–86.

Rinkevich, B., Lelker-Levav, T. and Goren, M. (1994a) Allorecognition/xenorecognition responses in *Botrylloides* (Ascidiacea) subpopulations from the Mediterranean coast of Israel. *J. Exp. Zool.* 270:302–313.

Rinkevich, B., Weissman, I.L. and Shapira, M. (1994b) Alloimmune hierarchies and stress-induced reversals in the resorption of chimeric protochordate colonies. *Proc. R. Soc. London (Biol.)*, 258:215–220.

Rinkevich, B., Shlemberg, Z. and Fishelson, L. (1995) Whole-body protochordate regeneration from totipotent blood cells. *Proc. Nat. Acad. Sci. U.S.A.* 92:7695–7699.

Sabbadin, A. (1962) Le basi genetiche della capacita di fusione fra colonie in *Botryllus schlosseri* (Ascidiacea). *Rend. Acad. Naz. Lincei*, 32:1031–1035.

Saito, Y. and Watanabe, H. (1982) Colony specificity in the compound ascidian, *Botryllus scalaris*. *Proc. Jpn. Acad.* 58B:105–108.

Satoh, N. and Jeffery, W.R. (1995) Chasing tails in ascidians: developmental insights into the origin and evolution of chordates. *Trends Genet.* 11:345–359.

Sawada, T., Zhang, J. and Cooper, E.L. (1993) Classification and characterization of hemocytes in *Styela clava*. *Biol. Bull.* 184:87–96.

Scofield, V.L., Schlumpberger, J.M., West, L.A. and Weissman, I.L. (1982) Protochordate allorecognition is controlled by MHC-like gene system. *Nature* 195:499–502.

Takahashi, H., Azumi, K. and Yokosawa, H. (1995) A novel membrane glycoprotein involved in ascidian hemocyte aggregation and phagocytosis. *Eur. J. Biochem.* 233:778–783.

Warr, G.W., Decker, J.M. and Mandel, D.D. (1977) Lymphocyte-like cells from the tunicate *Pyura stolonifera*: binding of lectins, morphological and functional studies. *Austr. J. Exp. Biol. Med.* 55:151–164.

Watters, D.J. and Van den brenk, A.L. (1993) Review article – toxins from ascidians. *Toxicon* 31:1349–1372.

Weissman, I.L., Saito, Y. and Rinkevich, B. (1990) Allorecognition histocompatibility in a protochordate species: Is the relationship to MHC semantic or structural? *Immunol. Rev.* 113:227–241.

Welsch, U. (1984) Urochordata. *In:* J. Bereiter-Hahn, A.G. Matoltsy and K.S. Richards (eds): *Biology of the Integument. 1. Invertebrates,* Springer-Verlag, Berlin, pp 800–816.

Wright, R.K. (1981) Urochordates. *In:* N.A. Ratcliffe and A.F. Rowley (eds): *Invertebrate Blood Cells, Vol. 2,* Academic Press, London, pp 565–625.

Yasuo, H. and Satoh, N. (1994) An ascidian homolog of the mouse brachyury (T) gene is expressed exclusively in notochord cells at the fate restricted stage. *Develop. Growth Differ.* 36:9–18.

Yund, P.O. and Feldgarden, M. (1992) Rapid proliferation of historecognition alleles in population of a colonian ascidian. *J. Exp. Zool.* 263:442–452.

Chordates/Vertebrates/Agnatha

Origin of the adaptive and anticipatory immune system within vertebrate subphylum should be considered in connection with the evolution of a whole body construction (the vertebrate body plan) and the emergence of new structures and organs forming the complex morphofunctional endowment of vertebrates. Only such an approach allows the true evaluation of the stepwise evolution of immune mechanisms which have arisen in main vertebrate taxa from cyclostomes through fish to mammals. It could be seen that sets of basic immune characters which had gradually emerged have to be shared later in evolution among the common predecessors from which modern vertebrate taxa have been derived.

Origin

The agnathans, commonly known as cyclostomes, are the most primitive species of the present-day vertebrates. Whereas the other taxa of jawless fish were extinct already in Paleozoic, the present two orders of cyclostomes, the *Myxiniformes* (hagfish) and the *Petromyzontiformes* (lampreys), have long and separate evolutionary histories. Even if their fossils may be found in the Carboniferous period and the representatives of both cyclostomate lineages seemingly have similar body design and anatomic features (eel-like body with no trace of dermal armor of their predecessors or paired fins, circular mouth with a complex toothed tongue), their origins and relationships are not yet clearly solved. Despite these common features, they differ substantially in many embryological developments, ontogeny (larvae are absent in myxinids), and in many other anatomical and non-morphological characters (e.g., the hagfish hemoglobin is dissimilar to those of lamprey and gnathostomean vertebrates) including their way of life (Løvtrup, 1977). Thus, contrary to the earlier hypotheses, all evidence suggests that these two groups separated a long time ago (Carroll, 1988). Recent cyclostomes are not considered to be the direct ancestors of jaw vertebrates. Palentologists regard the more advanced lampreys as a sister group of gnathostomes, but this is not valid for hagfish. Thus, these two cyclostomate orders do not form a monophyletic group (Forey and Janvier, 1993, Northcutt, 1996). This is important not only for paleontologists but also for evolutionary immunologists trying to discover the onsets of adaptive immunity.

The immunocompetent tissues and organs

The possible thymus equivalent lymphoid structures of pharyngeal region
The evidence supporting the presence of a thymus in myxinoids and lampreys is still absent. The thymus gland suggested in the Atlantic hagfish,

Myxine glutinose, by Müller (1845) and Stannius (1854) was shown to be pronefros (Papermaster et al., 1964; Kampmeier, 1969). Riviere et al. (1975) reported the presence of lymphoid cells in the muscle velar complex of Pacific hagfish, *Eptatetrus stoutii* and hypothesized that this structure is a protothymus, the direct phylogenetical precursor of the later thymic organ in the gnathostomes. However, due to the absence of information on the ontogeny of young hagfish, the question of the existence of thymus in myxinids remains open.

The observations documenting the thymus presence in lampreys are similarly controversial. The aggregates of lymphoid cells together with other blood elements in the branchial region of ammocoetes and have often been homologized to the true vertebrate thymus (Gerard, 1933; Finstad et al., 1964; Good et al., 1972). These accumulations are found in the vicinity of blood sinuses, located above and below the branchial clefts in epipharyngeal and hypopharyngeal folds, and in walls supporting branchial laminae. The whole pharyngeal region undergoes deep changes during metamorphosis which is of consequence in the transition of these structures to the lymphopoietic and blood-filtering tissues rather than that of thymus equivalent (Sterba, 1953; Percy and Potter, 1976; Yamaguchi et al., 1979; Page and Rowley, 1982; Ardavin et al., 1984; Ardavin and Zapata, 1988a).

The branchial region of ammocoetes is preferentially exposed to continual irritation caused by pathogens. Never ending external antigeneic stimuli and the necessity to defend the entrance into the internal environment could induce the development of foci of lymphoid cells which were specialized in filtration or trapping of particles, later forming a distinctive organ, the cavernous body. This is an enlarged blood sinus lined with endothelial phagocytic cells, containing lymphocytes and various differentiating stages of plasmacytes (Gerard, 1933; Nakao, 1978; Yamaguchi et al., 1979; Fujii, 1982; Ardavin and Zapata, 1988b). Due to the histoarchitecture and cell composition, the pharyngeal cavernous body shares more similarities with the medula of mammalian lymph nodes (Olah et al., 1975). The cavernous body together with other lymphoid pharyngeal structures may also have primary importance in the trapping of antigens included with food.

The gut-associated lymphoid structures considered to be equivalent to the spleen
There is no distinct spleen in hagfish. Only lymphohemopoietic cell aggregations accompanying the veins in the intestine submucosa forming about one-tenth of the gut wall volume (Grozdinski, 1926; Jordan and Speidel, 1930; Tanaka et al., 1981) remain a homology of vertebrates spleen with the connection of these aggregates with portal vein system. This conception, originally defended by numerous previous investigations (Jordan and Speidel, 1930; Finstad et al., 1964; Tomonaga et al., 1973) has already been

abandoned based on anatomical evidence (Tanaka et al., 1981). At present, it seems that the lymphomyeloid intestinal tissue has poietic rather than immune functions, thus resembling bone marrow of tetrapods (Zapata, 1983; Zapata and Cooper, 1990).

In lampreys, particularly in the larval stage of the ammocoete, the main lymphohemopoietic organ most often homologized to vertebrate spleen (Tanaka et al., 1981; Šíma and Slípka, 1995) or bone marrow (Percy and Potter, 1979; Fujii, 1981; Ardavin and Zapata, 1987) is a typhlosole. The accumulations of lymphoid cells in various developing stages, plasmacytes, erythrocytes, monocytes, and granulocytes are seen within the enlarged lamina propria of the invaginated gut epithelium enlarged in the form of the typhlosole along the anterior gut. The rich vascularization of the typhlosole resembles the vascular supply of mammalian bone marrow. Together with the reconstruction of the digestive tract during metamorphosis, a whole complex of lymphohemopoietic tissue of the typhlosole undergoes significant changes. There is a substantial increase of the connective tissue so that the typhlosole loses its capacity for renewal of blood cells. The lymphohemopoietic production moves to other tissues (see below).

Lymphoid tissues with the poietic capability supposed to be equivalent to bone marrow
Both in the intestinal wall (see above) and in the central mass of the hagfish pronephros (head kidney), lymphohemopoietic accumulations have been described (Holmgren, 1950). These accumulations are suspected to be equivalent to bone marrow. Head kidney is a paired organ derived from the wall of peritoneal cavity which has lost its original excretory function (Price, 1910). According to Willmer (1960), the pronephros may function as a primitive lymph node (presence of phagocytic cells) in cleaning the peritoneal fluid of foreign particles (Fänge, 1963). On the other hand, since ultrastructurally the pronephros contains lymphoid cells, plasmacytes, and various young stages of other blood cell types including erythroblasts, it is considered to fulfill the role of bone marrow (Zapata et al., 1984).

In petromyzontids, besides the typhlosole, the second tissue comparable to bone marrow is the nephric fold formed by the ammocoete opisthonephros (Jordan and Speidel, 1930; Sterba, 1962; Percy and Potter, 1976; Ardavin et al., 1984). The aggregates of immature and mature blood cells including plasmablasts, plasma cells, and large macrophages (Kilarski and Plytycz, 1981; Ardavin et al., 1984; Ardavin and Zapata, 1987) surrounded by connective tissue are situated intertubullary and in the adipose region of the opisthonephros. As in the case of typhlosole, the lymphohemopoietic capability here decreases significantly during metamorphosis. The nephrogenic tissue of posterior nephric fold transforms into the mature kidney (Ooi and Youson, 1979), where the blood cell formation within the adipose tissue continues (according to some authors to some degree only) at least to the end of metamorphose (Percy and Potter, 1977). In *P. marinus*, the adult ophistonephros constitutes important hematopoietic activity closely resembling that which could be seen in the adult teleostean kidney (Romer, 1971; Ardavin et al., 1974; Zapata, 1979).

The morphofunctional alterations and the loss of poietic microenvironment within the ophistonephros caused by the metamorphose are accompanied by the gradual transfer of lymphohemopoiesis to the superneural body which is non-functional until that stage. The supraneural body is situated in the dorsal sheath of the nerve cord. Composed of numerous lymphohemopoietic accumulations interlocated in adipose tissue, the histological architecture of this organ is the same as that of typhlosole and nephric fold of ammocoetes. In the adulthood of the lampreys, the supraneural body is the principal blood-forming organ classically regarded as a phylogenetic precursor of bone marrow, because it is predominantly granulopoietic in nature. However, besides immature and mature stages of all blood cell lineages, the abundant developing plasmacytes and lymphocytes could be seen in the suprarenal body, too (Ivanova-Berg and Sokolova, 1959; Piavis and Hiatt, 1971; Good et al., 1972; George and Beamish, 1974; Kelenyi and Larsen, 1976; Zapata, 1983; Ardavin et al., 1984). Here, in adult lampreys, the intensive proliferation of cells (Finstad and Good, 1964), increase in numbers of plasmacytes (Fujii, 1981), and appearance of antibody-producing cells (Hagen et al., 1983) after experimental immunization of adult lampreys is induced.

Other lymphohemopoietic tissues
In myxinids, there have been found lymphohemopoietic foci in the perivascular spaces of the gills, but it is not clear if true poietic activities take place there or if these aggregates originate secondarily by means of blood circulation and cell migration (Jordan and Speidel, 1930).

In larvae of petromyzontids, small groupings of lymphoid cells were found in the connective tissue of the liver localized in the portal region and interhepatic sinuses, and in the lamina propria of esophagial blood sinuses (Ardavin et al., 1984).

A survey of lymphohemopoietic organs and their supposed phylogenic morphofunctional equivalents is shown in Table 3.

Table 3. Possible phylogenetic equivalence of lymphohemopoietic structures of petromyzontids to main immunocompetent organs of tetrapods

Equivalent	Thymus	Spleen	Bone marrow	Lymph node
Ammocoetes	Branchial region	Typhlosole?	Typhlosole? Opisthonephros (nephric fold) Liver?	Cavernous body
Adult	?	?	Supraneural body Opisthonephros? Liver?	Pharyngeal region

Humoral immunity

Naturally occurring agglutinins (Boffa et al., 1967), bactericidins (Acton et al., 1969), and hemolysis (Day et al., 1970) have been found in agnathans. Most of the attention, however, has been focused on the complement system. As no clear evidence for a complement system in invertebrates has been reported to date (Farries and Atkinson, 1991), the cyclostomes as the most primitive living vertebrates are a natural target for studies of the evolution of complement system.

So far, only the alternative complement pathway has been found. Cyclostome complement is devoid of significant lytic action, but acts as the essential phagocytic factor (Nonaka et al., 1984). Fujii and Murakawa demonstrated that the serum of *Lampetra japonica* has lytic activity towards rabbit erythrocytes, which was inactivated after incubation with zymosan and was Mg^{2+} dependent (Fujii and Murakawa, 1981). Subsequent studies isolated and characterized an 192 kDa protein from the serum of the hagfish, *Eptatretus burgeri* (Fujii et al., 1992). This protein binds to zymosan particles, is composed of two distinct subunits of 115 kDa and 77 kDa, linked by disulfide bonds. The amino acid composition revealed a homology with the human C3 component of complement. Ishiguro et al. purified a three-subunit C3, which resembled mammalian C3, C4 and C5 components of complement and sequenced the corresponding cDNA clones (Ishiguro et al., 1992). So far it is unclear whether these two forms of C3 are generated by different processing of RNA or protein, or whether they are encoded by different genes. In a more recent paper, Fujii and coworkers described identification and isolation of the C3b fragment from the hagfish. The amino acid sequence analysis indicated that the hagfish C3b is analogous to mammalian C3b. This study also showed that the molecular features of degenerated C3 are altered during purification to generate a three-chain structure (Fujii et al., 1995).

Specific humoral immunity
Agnathans react to the antigenic stimulus such as sheep erythrocytes or *Brucella* antigen by formation of specific antibodies (Finstad and Good, 1964; Boffa et al., 1967; Fujii et al., 1979). Only IgM-like antibodies have been found so far with a sedimentation constant of 7S in hagfish (Thoenes and Hildemann, 1970) and 6.6S and 14S in lampreys (Marchalonis and Edelman, 1968). More recent isolation of immunoglobulins of the hagfish, *E. burgeri,* demonstrated that this immunoglobulin (MW of 150 kDa) was composed of both heavy and light chains (Kobayashi et al., 1985). Another study of the immunoglobulin isolated from the same species found the molecular weight being higher (210 kDa). Two distinct types of H chains have been found and these two chains, and L chain, appeared to be assembled in an unusual fashion (Hanley et al., 1990).

An interesting and stimulating observation is the study of Ishiguro et al. These authors isolated cDNA clones that encode a hagfish antibody, but found that the nucleotide sequences of the cDNA are not similar to mammalian immunoglobulins but are similar to complements C3, C4 and C5 (Ishiguro et al., 1992). Discoveries like this one place the general dogma that the adaptive immune system evolved in all vertebrates into the shadow of doubt (Andersson and Matsunaga, 1996).

Cellular immunity

Allograft rejection is well developed. Observed reactions such as inflammatory reaction, lymphocyte infiltration, capillary hemorrhage and pigment cell destruction are comparable to those seen in other vertebrates. First-set allografts showed a median survival time of 72 days, compared to 28 days for second-set allografts (Hildemann and Thoenes, 1969).

Leukocytes isolated from the Pacific hagfish, *E. stouti,* were found to respond to the chemotactic stimuli of mammalian C5a and LPS-activated plasma (Newton et al., 1994). Based on these findings, the authors hypothesize that the chemotactic stimulation may have been conserved during evolution and thus the chemotactic receptor on cyclostomes recognize mammalian chemoattractans.

A recognition molecule named complement-like protein, homologous to the C3, C4 and C5 vertebrate complement component, functions as an opsonin for phagocytosis (Raftos et al., 1992). Raison et al. showed that this opsonic activity can be decreased by preincubating leukocytes with an anti-leukocyte antibodies (Raison et al., 1994). In addition, antigen-activated recognition molecule can block the binding of the antibody to hagfish leukocytes. Immunoprecipitation studies indicated a 105 kDa cell surface protein expressed exclusively on phagocytic hagfish leukocytes specific for recognizing this complement-like protein.

References

Acton, R.T., Weinheimer, P.F., Hildemann, W.H. and Evans, E.E. (1969) Induced bactericidal response in the hagfish. *J. Bacteriol.* 99:626–628.

Andersson, E. and Matsunaga, T. (1996) Jaw, adaptive immunity and phylogeny of vertebrate antibody V_H gene family. *Res. Immunol.* 147:233–240.

Ardavin, C.F. and Zapata, A. (1987) Ultrastructure and changes during metamorphosis of the lympho-hemopoietic tissue of the larval anadromous sea lamprey *Petromyzon marinus. Dev. Comp. Immunol.* 11:79–93.

Ardavin, C.F. and Zapata, A. (1988a) The pharyngeal lymphoid aggregates of lampreys: a morpho-functional equivalent of the vertebrate thymus? *Thymus* 11:59–65.

Ardavin, C.F. and Zapata, A. (1988b) Lymphoid components in the branchial cavernous body of the ammocoete of *Petromyzon marinus. Acta Zool.* 69:23–28.

Ardavin, C.F., Gomariz, R.P., Barrutia, M.G., Fonfria, J. and Zapata. A. (1984) The lympho-hemopoietic organs of the anadromous lamprey, *Petromyzon marinus.* A comparative study. *Acta Zool.* 65:1–15.

Boffa, G.A., Fine, J.M., Drilhan, A. and Amouch, P. (1967) Immunoglobulins and transferrin in marine lamprey sera. *Nature* 214:700–702.

Carroll, R.L. (1988) *Vertebrate Paleontology and Evolution.* W.H.Freeman, New York.

Day, N. K. B., Gewurz, H., Johannsen, R., Finstad, J. and Good, R.A. (1970) Complement and complement-like activity in lower vertebrates and invertebrates. *J. Exp. Med.* 132:941–950.

Fänge, R. (1963) Structure and function of the excretory organs of myxinoids. *In:* A. Brodal and R. Fänge (eds): *The Biology of Myxine,* Universitetforslaget, Oslo, pp 516–590.

Farries, T.C. and Atkinson, J.P. (1991) Evolution of the complement system. *Immunol. Today* 12:295–300.

Finstad, J. and Good, R.A. (1964) The evolution of the immune response, III. Immunologic response in the lamprey. *J. Exp. Med.* 120:1151–1168.

Finstad, J., Papermaster, B. and Good, R.A. (1964) Evolution of the immune responses. II. Morphologic studies on the origin of the thymus and organized lymphoid tissue. *Lab. Invest.* 13: 490–512.

Forey, P. and Janvier, P. (1993) Agnathans and the origin of jawed vertebrates. *Nature* 361: 129-134.

Fujii, T. (1981) Antibody-enhanced phagocytosis of lamprey polymorphonuclear leucocytes against sheep erythrocytes. *Cell Tissue Res.* 219:41–51.

Fujii,T. (1982) Electron microscopy of the leucocytes of the typhlosole in ammocoetes, with special attention to the antibody-producing cells. *J. Morphol.* 173:87–100.

Fujii, T. and Murakawa, S. (1981) Immunity in lamprey. III. Occurrence of the complement-like activity. *Dev. Comp. Immunol.* 5:251–259.

Fujii, T., Nakagawa, H. and Murakawa, S. (1979) Immunity in lamprey. I. Production of haemolytic and haemogglutinating antibody to sheep red blod cells in Japanese lampreys. *Dev. Comp. Immunol.* 3:441–451.

Fujii, T., Nakamura, T., Sekizawa, A. and Tomonaga, S. (1992) Isolation and characterization of a protein from hagfish serum that is homologous to the third component of the mammalian complement system. *J. Immunol.* 148:117–123.

Fujii, T., Nakamura, T. and Tomonaga, S. (1995) Component C3 of hagfish complement has a unique structure: identification of native C3 and its degradation products. *Mol. Immunol.* 32: 633–642.

George, J.C. and Beamish, F.W.H. (1974) Haemocytology of the supraneural myeloid body in the sea lamprey during several phases of life cycle. *Can. J. Zool.* 52:1585–1589.

Gerard, P. (1933) Sur le systèmr sthrophsgocytaire chez l'ammocoète de la *Lampetra planeri* (Bloch). *Arch. Biol. Paris* 44:327–346.

Good, R.A., Finstad, J. and Litman, J. (1972) Immunology. *In:* M.W. Hardisty and I.C. Potter (eds): *The Biology of Lampreys,* Academic Press, London, pp 405–432.

Grozdinski, Z. (1926) Über das Blutgefäss-system von *Myxine glutinosa* L. *Bull. Acad. Pol. Sci. Cl. Sci. Math. Nat.* Series B:123–125.

Hagen, M., Filosa, M.F. and Youson, J.H. (1983) Immunocytochemical localization of antibody-producing cells in adult lamprey. *Immunol. Lett.* 6:87–92.

Hanley, P.J., Seppelt, I.M., Gooley, A.A., Hook, J.W. and Raison, R.L. (1990) Distinct Ig H chains in a primitive vertebrate, *Eptatretus stouti. J. Immunol.* 145:3823–3828.

Hildemann, W.H. and Thoenes, G.H. (1969) Immunological responses of Pacific hagfish. I. Skin transplantation immunity. *Transplantation* 7:506–521.

Holmgrem, N. (1950) On the pronephros and the blood in *Myxine glutinosa. Acta Zool.* 31: 233–348.

Ishiguro, H., Kobayashi, K., Suzuki, M., Titani, K., Tomonaga, S. and Kurosawa, Y. (1992) Isolation of a hagfish gene that encodes a complement component. *EMBO J.* 11:829–837.

Ivanova-Berg, M.M. and Sokolova, M.M. (1959) Seasonal changes in the blood of the river lamprey (*Lampetra fluviatilis* L.). *Vop. Ikthtiol.* 13:156–162.

Jordan, H.E. and Speidel, C.C. (1930) Blood formation in cyclostomes. *Am. J. Anat.* 46: 355–391.

Kampmeier, O.F. (1969) *Evolution and Comparative Morphology of the Lymphatic System.* Charles C. Thomas, Springfield.

Kelenyi, G. and Larsen, L.O. (1976) The haematopoietic supraneural organ of adult sexually immature river lampreys (*Lampetra fluviatilis* L. Gray) with particular reference to azurophilic leucocytes. *Acta Biol. Acad. Sci. Hung.* 27:45–56.

Kilarski, W. and Plytycz, B. (1981) The presence of plasma cells in the lamprey *(Agnatha). Dev. Comp. Immunol.* 5:361–366.

Kobayashi, K., Tomonaga, S. and Hagiwara, K. (1985) Isolation and characterization of immuno-globulin of hagfish, *Eptatretus burgeri,* a primitive vertebrate. *Mol. Immunol.* 22: 1091–1097.

Løvtrup, S. (1977) *The Phylogeny of Vertebrata.* John Wiley and Sons, London.

Marchalonis, J.J. and Edelman, G.M. (1968) Phylogenetic origins of antibody structure. III. Antibodies in the primary immune response of the sea lamprey, *Petromyzon marinus. J. Exp. Med.* 127:891–914

Müller, J. (1845) *Vergleichende Anatomie der Myxinoiden. III. Über das Gefäss-System.* Abhandl. Preuss. Akad. Wiss., Berlin.

Nakao, T. (1978) An electron microscopic study of the cavernous bodies in the lamprey gill fila-ment. *Am. J. Anat.* 151:316–336.

Newton, R.A., Raftos, D.A., Raison, R.L. and Geczy, C.L. (1994) Chemotactic responses of hagfish (*Vertebrata, Agnatha*) leucocytes. *Dev. Comp. Immunol.* 18:295–303.

Nonaka, M. and Takahashi, M. (1992) Complete complementary DNA sequence of the third component of complement of lamprey: implication for the evolution of thioester containing proteins. *J. Immunol.* 148:3290–3295.

Nonaka, M., Fujii, T., Kaidoh, T., Natsuume-Sakai, S., Yamaguchi, N. and Takahashi, M. (1984) Purification of a lamprey complement protein homologous to the third component of the mammalian complement system. *J. Immunol.* 133:3242–3249.

Northcutt, R.G. (1996) The agnathan ark: the origin of craniate brains. *Brain Behav. Evol.* 48: 237–247.

Olah, I., Rohlich. P. and Toro, I. (1975) *Ultrastructure of lymphoid Organs.* Masson, Paris.

Ooi, E.C. and Youson, J.H. (1979) Regression of the larval opisthonephros during metamor-phosis of the sea lamprey *Petromyzon marinus* L. *Am. J. Anat.* 154:57–59.

Papermaster, B.W., Condie, R.M., Finstad, J. and Good, R.A. (1964) Phylogenetic development of adaptive immunologic responsiveness in vertebrates. *J. Exp. Med.* 119:105–130.

Page, M. and Rowley, A.F. (1982) A morphological study of pharyngeal lymphoid accumula-tion in larval lampreys. *Dev. Comp. Immunol. Suppl.* 2:35–40.

Percy, L. R. and Potter, I.C. (1976) Blood cell formation in the river lamprey, *Lampetra fluvia-tilis. J. Zool.* 178:319–340.

Percy, L.R. and Potter, I.C. (1977) Changes in haemopoietic sites during metamorphosis of the lampreys *Lampetra fluviatilis* and *Lampetra planeri. J. Zool.* 183:111–123.

Percy, L.R. and Potter, I.C. (1979) The intestinal blood circulation in the river lamprey, *Lampe-tra fluviatilis. J. Zool.* 187:415–431.

Piavis, G.W. and Hiatt, J.L. (1971) Blood cell lineage in the sea lamprey, *Petromyzon marinus (Pisces: Petromyzontidae). Copeia* 4:722–728.

Price, G.C. (1910) The structure and function of the adult head kidney of *Bdellostoma stoutii. J. Exp. Zool.* 9:848–864.

Raftos, D.A., Hook, J.W. and Raison, R.L. (1992) Complement-like protein from the phyloge-netically primitive vertebrate, *Eptatretus stouti,* is a humoral opsonin. *Comp. Bioch. Physiol.* 103:379–385.

Raison, R.L., Coverley, J., Hook, J.W., Towns, P., Weston, K.M. and Raftos, D.A. (1994) A cell-surface opsonic receptor on leucocytes from the phylogenetically primitive vertebrate, *Epta-tretus stouti. Immunol. Cell Biol.* 72:326–332.

Riviere, H.B., Cooper, E.L., Reddy, A.L. and Hildemann, W.H. (1975) In search of the hagfish thymus. *Am. Zool.* 15:39–49.

Romer, A.S. (1971) *The Vertebrate Body.* Saunders, Philadelphia.

Síma, P. and Slípka, J. (1995) The spleen and its coelomic and enteric history, *In:* J. Mestecky, M.W. Russell, S. Jackson, S.M. Michalek, H. Tlaskalova-Hogenova and J. Sterzl (eds): *Advances in Mucosal Immunology, Part A,* Plenum Publ. Corp., New York, pp 331–334.

Stannius, H. (1854) *Handbuch der Zootomie,* Berlin.

Sterba, G. (1953) Die Physiologie und Histogenese der Schildrüse und des Thymus beim Bach-neunauge (*Lampetra planeri* Bloch = *Petromyzon planeri* Bloch) als Grundlage phylogeneti-scher Studien über die Evolution der innensekretorischen Kiemenderivate nebst eingehenden Mitteilungen über die Bionomie der Bachneunaugen und morphologisch-physiologischen Untersuchungen über den Kiemendarm. *Wiss. Zeitschr. F Schiller Univ. Jena* 239: 239–298.

Sterba, G. (1962) Die Neunaugen *(Petromyzondtidae). Hand. Binnenfish. Mittl.* 3B:363–351.

Tanaka, Y., Saito, Y. and Gotoh, H. (1981) Vascular architecture and intestinal hemopoietic nests of two cyclostomes, *Eptatretus burgeri* and ammocoetes of *Entosphenus reissneri:* A com-parative morphological study. *J. Morphol.* 170:71–93.

Thoenes, G.H. and Hildemann, W.H. (1970) Immunological responses of Pacific hagfish. II. Serum antibody production to soluble antigens. *In:* J. Sterzl and I. Riha (eds): *Developmental Aspects of Antibody Formation and Structure,* Academia, Prague, pp 711–722.

Willmer, E.N. (1960) *Cytology and Evolution.* Academic Press, New York.

Yamaguchi, K., Tomonaga, S., Ihara, K. and Awaya, K. (1979) Electron microscopic study of phagocytic lining cells in the cavernous body of the lamprey gill. *J. Electron Microsc.* 28:106–116.

Zapata, A. (1979) Ultrastructural study of the teleost fish kidney. *Dev. Comp. Immunol.* 3:55–65.

Zapata, A. (1983) Phylogeny of the fish immune system. *Bull. Inst. Pasteur* 81:165–186.

Zapata, A. and Cooper, E.L. (1990) *The Immune System: Comparative Histophysiology.* John Wiley and Sons, Chichester.

Zapata, A., Fänge, R., Mattison, A. and Villena, A. (1984) Plasma cells in adult Atlantic hagfish *Myxine glutinosa. Cell Tissue Res.* 235:691–693.

Chordates/Vertebrates/Chondrichthyes

"One of the great events of revolutions in the history of vertebrates was the appearance of the jaws. The importance of this evolutionary development can hardly be overestimated, for it opened to the vertebrates new lines of adaptation and new possibilities for evolutionary advancement that expanded immeasurably the potentialities of these animals... The appearance of the jaws in vertebrates was brought about by a transformation of anatomical elements that originally had performed a function quite different from the function of food gathering." (Colbert, 1980).

Actually, this quotation may be particularly true with respect to the emergence of adaptive immune mechanisms. Matsunaga and Andersson (1994) have proposed: "The new ability to bite and swallow food by animals with the jaw would have caused increased frequency of physical injuries in the wall of digestive tract (esophagus, stomach and intestine) of those primitive jawed fishes, which eventually led to the development of adaptive immunity." It could be easily imagined that these animals explored all available mechanisms which they had inherited from their ancestors for watching and defending their digestive tract. The first sign of the significant role of gut-associated lymphoid tissue (GALT) in the adaptive immunity could be seen already in agnathans (see Section on *Agnatha*). Thus, it is understandable that from the evolutionary stage of the first jawed animals, the GALT has gained its main immune importance. Because of that, it is the largest immune organ in all vertebrates.

Origins

The members of the class *Chondrichthyes*, commonly called the cartilaginous fish, are divided into two main assemblages, the elasmobranch (*Elasmobranchii*) comprising modern sharks, skates and rays, and the holosteans (*Holocephali*), the chimaeras. For details on taxonomy of this class, see (Jarvik, 1980; Carroll, 1988; Šíma and Větvička, 1990).

First fish-like gnathostomean remains are known from the Silurian, almost 50–100 million years after ancestors of agnathans. The chondrichthyans have undergone two major adaptive radiations during their evolution. The first one occurred during Paleozoic Period, but these archaic animals became extinct by the end of Triassic Period. During Jurassic Peroid, the second adaptive radiation gave rise to an ancestral assemblage, the ctenacanthids, from which the predecessors of neoselanchians have evolved. Despite some different features proper to the holocephalans, it is today believed that they evolved separately from their common ancestors with elasmobranchs. Both assemblages of chondrichnthyans are often considered to be rather primitive animals in comparison to the osteichthyans because of their cartilaginous endoskeleton. This opinion is no longer tenable. From the fossil findings point of view, this typical mark has evolved secondarily from ossified skeleton of their ancestors.

The immunocompetent tissues and organs of chondrichthyans

The thymus

A tendency to the evolution of an organized immunocompetent tissue in the branchial region giving later to the rise of the thymus could be followed from the beginning of first chordates up to gnathostomes where it has became a distinct organ playing a fundamental, highly specialized role in immunogenesis (Slípka and Síma, 1989). The ancestral immune structures began to evolve at first in close connection with the rostral part of the digestive tube, within the branchial region. In parallel with their evolution, an excretory nephridial system developed which in cephalochordates and cyclostomes joined the circulatory system. The nephridia, originally maintaining the osmotic balance, gradually started to serve for removal of antigens. In higher steps of evolution, the more organized tissue of pro- and mesonephros became lymphohemopoietic tissue (Willmer, 1970). On the evolutionary level of gnathostomes, the poietic potention moved into neighboring mesenteries, in which the liver developed. The hepatolienal type of lymphohemopoiesis became a main source of all blood elements. The excretory type of the antigen removal was structurally improved and shifted in the cranio-caudal direction, too. The nephridia of *Amphioxus* remained still opened in branchial clefts but in all true vertebrates the pronephros became situated caudally behind the last gill slit, out of the branchial region. The branchial clefts were in this way liberated for other functions. They gradually entered the immune processes and their epithelium became a primordium of the central immune organ, the thymus (Phisalix, 1885; Beard, 1902; Salkind, 1915).

The thymus as a well-developed distinct organ appears for the first time in chondrichthyans. It develops from the dorsal parts of all clefts, and even in some primitive sharks (*Heptanchus*) a connection of thymus with gill slits via epithelial ducts suggests its primordial excretory character resembling the "branchionephros" in *Amphioxus*. Remnants of these ducts are recapitulated even during ontogeny of human thymus (Slípka, 1979). Macroscopically, the elasmobranchian thymus is a multilobulated organ surrounded by a connective tissue capsule situated near the gill region. Thymic histological organization into clearly distinguishable cortex and medulla is similar to those known in the thymus of higher vertebrates with the possible exception of *Scyliorhinus canicula* (Pulsford et al., 1984). True Hassall's bodies are lacking, even if some authors occasionally cited their presence in the thymus of some chondrichthyan species (Beard, 1902; Good et al., 1966). It seems rather that the structures resembling Hassall's bodies are masses of fibrous material or aggregates of medullary reticuloepithelial cells (Fänge and Sundell, 1969) and degenerative epithelial cysts. Ultrastructural studies describe chondrostean thymus as practically lymphoid with the exception of chimaerids where the thymus appears to be a hemopoietic organ (Fujita, 1963). Thymic lymphocytes of various sizes, often in mitosis, and lymphoblasts are mainly seen in the cortex reticulocyte network, whereas epithelial cells prevail in the medulla. Beside active macrophages responsible for phagocytosis, the phagocytic activity of epithelial cells in the thymuses of *Torpedo marmorata* and *Raja clavata* were found (Zapata, 1980a). Ig-positive lymphocytes were described in the thymus of *Raja naevus* (Ellis and Parkhouse, 1975), but not in *R. kenojei* (Tomonaga et al., 1984). Other cell types such as the myoid, neuroendocrine-like cells and other phagocytic elements occur mainly in the thymus medulla (Zapata, 1980b; Mattison and Fänge, 1986; Chibá et al., 1989; Zapata and Cooper, 1990; Zapata et al., 1996). Apart from a row of histological studies, there is very little information available on the ontogenesis of the thymus. Navarro (1987) and Lloyd-Evans (1993) in *S. canicula* have discriminated the thymus primordium in the dorsal part of pharyngeal epithelium on both sides of the gill arches of 2-month old embryos. As to the thymic involution with age, at least in some species the thymus is not constantly present (Fänge, 1977; Fänge and Mattisson, 1981).

The spleen
In the evolutionary line, the first spleen as a distinct organ emerges in the sharks as an elongated, anatomically highly organized structure situated in the dorsal mesentery. It is supplied by arteria lienalis leading blood from aorta just next to the arteria mesenterica superior which lies in the shallow grove of the spiral fold. A close anatomical relationship with spiral fold could suggest some evolutionary morphofunctional connections of agnathan typhlosole → chondrichthyan spiral valve → spleen (Síma and Slípka, 1995). Undoubtedly, the structural history of spleen and its function were originally different, and both have changed during the ages. It began as a structure which served blood filtration and phagocytosis, then it became more active in hemopoiesis, and finally it acquired immune function.

From comparative aspects, the development and structure of the spleen have been studied in many species of cartilaginous fish in monographies of Tischendorf (1969), Zapata and Cooper (1990), and recently in the review written by Zapata et al. (1996). The chondrichthyan spleen is a well-developed organ with the white pulp composed of large masses of lymphoid cells and developing plasma cells around larger blood vessels with lymphoipoietic capability, and the red pulp in which both the production and the degradation of erythrocytes proceed and the thrombocytes are formed (Zapata 1980b; Pulsdorf et al., 1982). Sometimes, the distinction between the white and the red pulp is less visible. Similar spleen activities have also been described for a holocephalan *Chimaera monstrosa* (Fänge and Sundell, 1969). The spleen in elasmobranchs is a main site where antibodies are formed (Morrow, 1978; Tomonaga et al., 1984, 1985; Pulsdorf and Zapata, 1989). After splenectomy the antibody production is not influenced and could be compensated in other lymphoid organs (Ferren, 1967). In endothermic vertebrates, the immune response is connected with the germinal centers which seem to be absent in elasmobranchs. Similarly, the presence of melanomacrophage centers known in teleostean fish, often considered as evolutionary predecessors of germinal centers, has not been unequivocally proven in cartilaginous fish (Zapata et al., 1996).

The gut-associated lymphoid tissue (GALT)
In contrast to the agnathan gut epithelia and typhlosole in which first simple aggregations of lymphoid elements could be seen, both the true small lymphoid nodule-like or massive accumulations of granular, macrophage, lymphoid, and plasma cells including immunoglobulin-forming cells have been found in various species of elasmobranchs in the spiral valve or duodenal lamina propria and gut epithelium (Tomonaga et al., 1986; for review see Zapata et al., 1996).

The immunocompetent tissues and organs of chondrichthyans comparable to the bone marrow
In chondrichthyans no structures homological to the bone marrow known in higher vertebrates have been identified *in sensu stricto*. The tissues where the primordial blood polypotent stem cells find suitable microenvironment to give rise to blood cell lines are found in the walls of upper part of the digestive tract, outer region of the gonads, whereas the lymphohemopoietic centers in other locations such as in the kidney or meninges could be found in some species only (Zapata et al., 1995).

The Leydig's organ
This lymphomyeloid tissue with poietic capacity located in the walls of the esophagus was named after Leydig, who first defined it as a "lymph node-like" structure (Leydig, 1857), even if this organ has been known since 1685, and later was considered to be a salivary gland (Chiaje, 1840). Histologically, there are no substantial differences in Leydig's organ among various species of elasmobranchs. It consists of a series of patches of lymphomyeloid tissue in the mucosa (Maximov, 1923) arranged as two bodies ventrally and dorsally between the esophageal mucosa and muscularis, extending from the oral cavity through the whole length of the esophagus, up to the stomach (Pilliet, 1890; Petersen, 1908; Zapata, 1981). It includes lymphoid cells in various developing stages, all types of granulocytes except basophils (which seem to be completely lacking in chondrichthyans), and reticulocytes. Normally, the organ is concerned only with the production of leukocytes, but in the basking shark, *Cetorhinus maximus,* active erythropoiesis was reported (Matthews, 1950), and after splenectomy, it may assume the compensatory function of producing

erythrocytes and thrombocytes as well (Fänge et al., 1975). Apart from the hemocytopoietic function of Leydig's organ, almost nothing is known of its possible immune functions. Leydig's organ is present in a prevailing number of elasmobranch species studied. It could be sometimes underdeveloped, forming only residual lymphohemopoietic tissue, or in several species could be fully absent. In that case, the spleen could be relatively diminished and the epigonal organ is then enlarged (Fänge, 1977, 1982).

Epigonal organs
Most of that what was stated about Leydig's organ is valid also for epigonal organs. Apart from the different anatomical position of epigonal organs, their structure and cell composition are very similar, if not the same as in Leydig's organ (see above). Both organs have similar histoarchitecture, lymphohemopoietic function, and cellular composition. There is reciprocal connection between the Leydig's and epigonal organs: if one organ is lacking, the second becomes more voluminous and compensates the poietic functions of the missing one (Fänge, 1984). The development and the presence of epigonal organs are variably dependent on the species. They may be situated as huge organs extending from the liver to rectal gland, bilaterally or unilaterally. In other species they may be intermingled like an irregular mass in the mesenterial folds of the gonads, or form only the posterior part of the gonad (Fänge, 1987). Tomonaga et al. (1985) have revealed some immunoglobulin forming cells in these organs, but nothing more is known about their role in chondrichthyan immunity. Some authors suggest that due to specific microanatomy (vascular sinusoids), these organs could have a filtration function where contact of immunocompetent cells with antigens could occur (Zapata and Cooper, 1990). Together with Leydig's and possibly other lymphomyeloid masses localized in the kidneys, digestive tract, and mainly throughout the cranial region, the epigonal organs actually represent the structures functionally comparable to bone marrow of higher vertebrates.

The lymphomyeloid structures of cranial region
The lymphomyeloid structures exerting poietic activities associated with the primitive meninges, choriod plexuses, and within the central nervous system are considered to be equivalent of bone marrow. They have been described in some elasmobranch species (Vialli, 1933; Chibá et al., 1988), holocephalans (Stahl, 1967; Mattison and Fänge, 1986; Mattisson et al., 1990), less advanced osteichthyan fish (Chandler, 1911; Scharrer, 1944), urodelans (Dempster, 1930; Sano and Imai, 1961), and even in early human embryos (Kappers, 1958). In the holocephalan *Chimaera monstrosa* which otherwise appears to lack well-developed lymphoid structures in other organs possible with the exception of spiral valve (Jacobshagen, 1915), the lymphohemopoietic tissue resembling the epigonal or Leydig's organs has been identified in the orbital and preorbital canal of the cranial cartilage,

and in a depression in the basis cranii in 1851 by Leydig (Leydig, 1851; 1857, Holmgren, 1942; Stanley, 1963; Zapata, 1983). Recently, the existence of macrophage-lymphocyte cells accumulations in the hypothalamic ventricle of some specimens of the stingray *Dasyatis akajei*, cloudy dogfish *S. torazame*, smoot dogfish *Triakis scyllia*, and gummy shark *Mustelus manazo* has been reported (Torroba et al., 1995). The meningeal lymphohemopoietic tissue of the stingray is characteristic for every local microenvironment (Chibá et al., 1988). In telencephalon, it contains mainly developing granuloid leukocytes localized between collagenous supporting fibers, in diencephalon it consists predominantly of lymphoid cells, and in mesencephalon the macrophage-lymphocyte clusters are typical and resemble similar pattern found in some mammals (Nielsen et al., 1974). The meningeal lymphohemopoietic structures are probably more engaged in the immune reactions. Recently, Torroba et al. (1995) have documented the presence of macrophages, lymphocytes, and plasma cells in brain ventricles where they may form clusters or organized lymphoid tissue along the large blood vessels, but the cell elements never penetrate the brain parenchyma.

The kidney

In comparison to lymphomyeloid tissue found in the kidney of holostean and chondrostean fish, and in the pronephros (head kidney) of teleosteans, most investigators are convinced of its existence in elasmobranch kidney, although it has been documented in a few species (Schneider, 1897; Policard, 1902; Maximow, 1910; Lacy and Reale, 1985). On the other hand, at least in some species studied, the embryonic kidney is the first place where the lymphohemopoietic activity occurs during ontogeny (Navarro, 1987; Hart et al., 1988; Lloyd-Evans, 1993). Moreover, the Ig expression has been detected on the surface of most lymphoid cells (Lloyd-Evans, 1993). Later in development, the poietic activity declines so that in mature animals it disappears completely. No lymphohemopoiesis seems to take place in the urogenital system of chimaerids (Stanley, 1963).

Development of lymphohemopoiesis in elasmobranchs

When the development of the immune organs has been studied in embryonic and post-hatch dogfish, *S. canicula*, it has been found that the major feature is the succession of lymphoid tissues (Hart et al., 1986). At a stage of external gills development, the primordia of the thymus, the spiral valve and the kidney appear, but no spleen, Leydig's or epigonal organs. The liver was the first tissue with Ig-positive cells, followed by the interstitial kidney and spiral valve. The lymphoid infiltration in the spleen, thymus and Leydig's organ together with epigonal organs appears later at the stage when a young shark is ventilated by sea water. The hematopoietic/lympho-

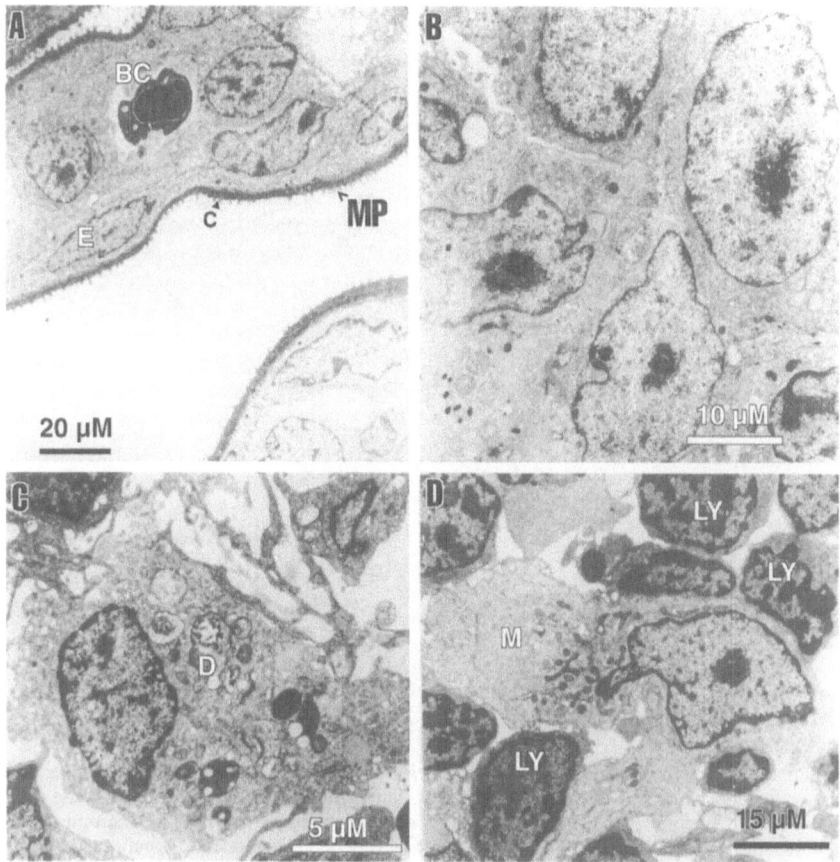

Figure 5. A) Electron micrograph of the thymus at pre-hatch stage (8 months) showing two adjacent lobes of thymic tissue bound by a one-cell-thick capsule (C) of endothelial cells (E) and distinctive microprojection (MP). The lobes are permeated by a network of capillaries (BC). B) Large lymphoblasts (LB) are a characteristic of the thymus at all stages. C) A macrophage showing large nucleus and an extensive cytoplasm with debris (D). They are associated with the medullary regions of the thymus tissue at all stages. D) Peripheral region of the thymus with abundant lymphocytes (LY), which are often in close association with other cell types such as macrophages (M). (From Lloyd-Evans, 1993, with permission.)

id nature of kidney and thymus disappeared at post-hatch (Lloyd-Evans, 1993). By the time when animals receive massive exposure to water-born antigens, the structural development of most of the lymphomyeloid tissues is well advanced (see Figs 5 and 6).

The survey of the development of lymphohematopoietic structures found in *Scyliorhinus canicula* and suggested phylogenic equivalence to main immunocompetent organs of tetrapods is summarized in Table 4.

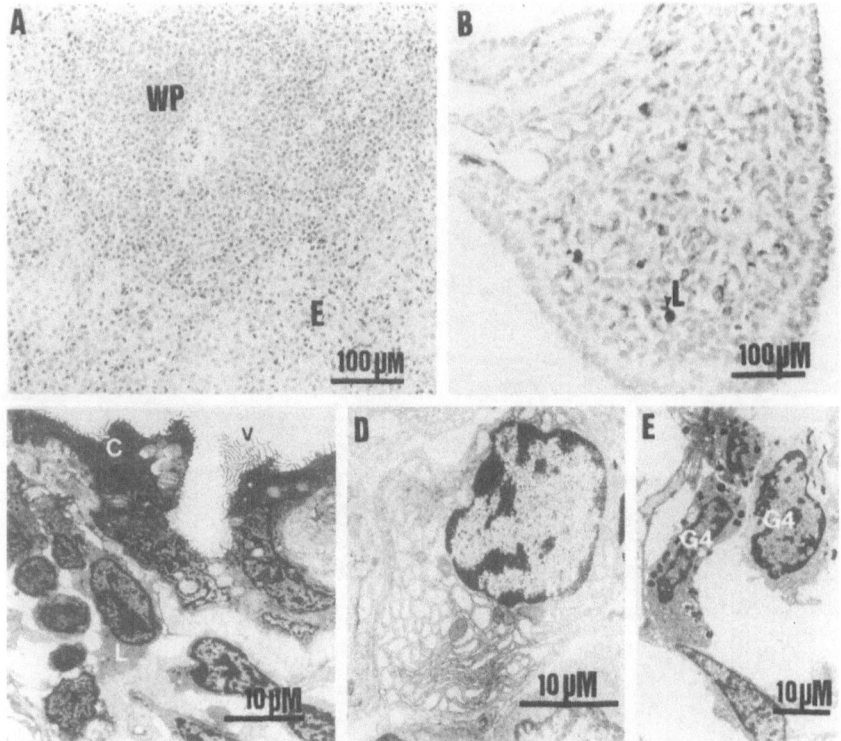

Figure 6. A) Longitudinal section of spleen at pre-hatch stage (8 months) showing clear division into white pulp (WP) and red pulp (E) areas. The white pulp is characterised by dense accumulations of lymphocytes and the red pulp contains developing erythrocytes, thrombocytes, and leukocytes. B) Spleen at external gill stage (4 months) remains undifferentiated and contains a few Ig⁺ cells. C) An electron micrograph of the spleen at post-hatch showing the microvillar outer border (V) of capsule cells (C). Leukocytes (L) are found immediately below. D) Plasma cell in the spleen with distended rough endoplasmic reticulum and large nucleus. E) Type IV granulocytes (G4) showing elongated structure and granules. (From Lloyd-Evans, 1993, with permission.)

Humoral immunity

Lemon shark and nurse shark serum contain natural agglutinins to chicken and sheep red blood cells (Leslie and Clem, 1970).

The particularly primitive sharks are the nurse sharks. These species were used for studies of the complement system. The serum has very high levels of lytic activity against various unsensitized erythrocytes (Gigli and Austen, 1971). This activity can be blocked by absorption with erythrocyte stromata, but the absorbed serum can still serve as a complement source when sensitized erythrocytes are used, suggesting the presence of the complement-dependent antibody system (Legler and Evans, 1967). So far, six functionally pure complement components have been described – C1, C2,

Table 4. The survey of the development of lymphohematopoietic structures found in *Scyliorhinus canicula* and suggested phylogenic equivalence to main immunocompetent organs of tetrapods.

	Thymus	Spleen	Bone marrow	GALT
Equivalent	Thymus	Spleen	Leydig or epigonal organs; kidney	Spiral valve
Prehatching	Epithelial Primordium In pharynx			
	Lymphoid Colonies	Spleen Primordium	Renal Primordium	Lymphocytes Macrophages in 1. propria
	Vascularisation	Erythrogranulo-cytopoiesis	Myeloid cells in oesophagal submucosa and medulla of gonads granulopoiesis	Adult spiral valve Lymphocyte Clusters in lamina propria
	Differentiation of lymphocytes and macrophages	Vascularisation	Lymphopoiesis in embryonal kidney; lymphoid cells in Leydig and epigonal organs	Intraepithelial lymphocytes Macrophages
	Cortex and medulla	Lymphocytes Macrophages in red pulp	Decrease of poietic activity in kidney	
Adult		Melono-Macrophages	No activity in adult kidney	Granulocytes Plasma cells in spiral valve

C3, C4, C8, and C9 (Jensen et al., 1981). Cobra venom factor treatment of the shark serum failed to induce lysis of erythrocytes, even though the hemolytic activity of the serum was highly reduced, which suggests the activation of an alternative complement pathway (Day et al., 1970).

In contrary to these findings, only negligible levels of lytic activity have been found in rays (Legler and Evans, 1967).

An interesting observation was the discovery of a squalamine, an aminosterol antibiotic from the shark *Squalus acanthia* (Moore et al., 1993). This antibiotic exhibits potent bactericidal activity against both Gram-negative and Gram-positive bacteria, is fungicidal and induces osmotic lysis of protozoa.

Specific humoral immunity
Antibodies in the Chondrichthyes were studied in the 1960s. Using the smooth dogfish, *Mustelus canis*, and the lemon shark, two IgM fractions, 7S and 17S were found (Marchalonis and Edelman, 1965, 1966; Clem and Small, 1967; Sigel, 1974). Both types of immunoglobulins had light and heavy chains.

When the antibody response was evaluated, antigeneically identical 7S and 17S IgM were produced. Further studies showed that the 17S molecule is a pentamer of the 7S form. The 17S molecule is the first one to be secreted, the 7S levels gradually increase and around 1 year after immunization the levels of both fractions are equal (Suran et al., 1967; Goodman et al.; 1970). On the other hand, Voss and Sigel (1971) demonstrated an 19S to 7S to 19S immunoglobulin cycle in nurse sharks. Both proteins have identical specificity and similar binding constant, but the 19S pentamers form more stable complexes with antigen. For a detailed study of shark natural antibodies, see (Marcholonis et al., 1993).

All older studies of shark immunoglobulin agreed on one point – there is only one class of antibodies. However, the recent paper by Bernstein et al., used a rapid amplification of cDNA ends and a highly conserved constant region consensus amino acid sequence and isolated a new immunoglobulin class from the sandbar shark, *Carcharhinus plumbeus* (Bernstein et al., 1996). This immunoglobulin, named IgW, in its secreted form consists of 782 amino acids and is expressed in both the spleen and the thymus. It most closely resembles μ chains of the skates and human and a new putative antigen-binding molecule (NAR) isolated from the nurse shark (Greenberg et al., 1995). Comparison of the sequences of IgW V and C domains shows homology greater than that found among V_H and C_μ, suggesting that IgW may retain features of the primordial immunoglobulin (Bernstein et al., 1996). Another novel immunoglobulin class has been recently demonstrated in nurse shark. This molecule has an amino-terminal V domain, followed by six C-1 set domains, and is ending in a carboxy-terminal tail typical of secreted IgM, IgA and NAR (Greenberg et al., 1996). The two C domains are orthologous to IgR in the skates, the last four domains are homologous to the carboxy-terminal four domains of NAR. These data suggest that the arsenal of secreted antigen receptors in sharks is greater than previously supposed. In addition, the current dogma that IgM is the primordial isotype, seems to be on shaky ground.

Using a model of the most primitive shark, the frill shark, *Chlamydoselachus anguineus*, Kobayashi et al. identified a second type of immunoglobulin (Kobayashi et al., 1992). They found three molecular forms of immunoglobulins, pentamer, dimer, and monomer. The pentamer with a MW of 900 kDa is an IgM, based on similarities to mammalian IgM in both molecular form and H chain molecular weight. The dimer and monomer with MW of 300 kDa and 150 kDa, respectively, have identical H chains of 45-50 kDa (compared to 68 kDa of the pentamer), and light chains identical to those of the pentamer. The H chains were synthesized by different plasma cells.

A different situation exists in spiny rasp skate, *R. kenojei*, and the Auletian skate, *Bathyraja aleutica*. An 18S IgM molecule with MW of 840 kDa and an 8.9S IgR molecule with MW of 320 kDa. This immunoglobulin was found to be secreted by the plasma cells in a different way from those producing IgM (Kobayashi et al., 1984, 1985, Tomonaga et al., 1986).

Based on these facts, it can be concluded that the *Rajiformes* of the cartilaginous fish have a more evolved immunoglobulin system than that of bony fish. However, the latest findings seem to suggest that at least some sharks also developed more types of immunoglobulins.

The prototypic chondrychthian immunoglobulin light chain type isolated from horned shark, *Heterodontus francisci*, has a clustered organization in which variable (V), joining (J), and constant (C) elements are in relatively close VJC linkage. When the second light chain types from *Heterodontus* and *R. erinacea* were tested, it was found that V, J, and C regions are again arranged in closely linked clusters (Rast and Litman, 1994; Rast et al., 1994). However, by comparison to mammals, the antibody repertoire of sharks is limited, due to a lack of combinatorial association between the V, D and J gene segments (Parham, 1995). Generation of immunoglobulin light chain gene diversity in *R. erinacea* is not associated with somatic rearrangement, which is an exception to a central paradigm of B cell immunity (Anderson et al., 1995). Adding to the restriction imposed on the IgM response is the absence of affinity maturation through somatic hypermutation (Makela and Litman, 1986). IgM may comprise as much as 50% of serum proteins in the shark. Comparison of the natural antibodies in sharks and humans revealed a high cross-reactivity of shark IgM antibodies (Marchalonis et al., 1993) comparable to that shown by human polyspecific IgM autoantibodies produced by CD5$^+$ B cells.

Schluter et al. compared the constant region gene sequence between shark and mammalian immunoglobulin light chains (1990) and found homology. The experiments showed that shark light chains exhibit heterogeneity at both the gene and protein level, but that the constant regions of these chains can be identified as homologs of mammalian λ chains and that the evolutionary conservation has occurred in V region sequences ranging from elasmobranchs to man. Similarly, canonical structures occurring most commonly in hypervariable regions are the same in cartilaginous fishes, mice and humans, suggesting that these structures arose very early in the evolution of the immune system (Barre et al., 1994).

Cellular immunity

Three types of cells are involved in phagocytosis: tissue macrophages, acidophilic granulocytes, and eosinophilic granulocytes (Hyder et al., 1983). When the clearance of a T2 bacteriophage was studied, the elimination was found to be so fast, that by day 2 the amount of circulating bacteriophage was reduced more than 1000 times (Sigel et al., 1968). Other studies showed these cells differ in ability to internalize different types of prey. Phagocytosis of target cells by neutrophils is shown in Figure 7.

McKinney described the presence of Fc receptors for IgM on shark leukocytes (Haynes and McKinney, 1991), but not macrophages (McKinney

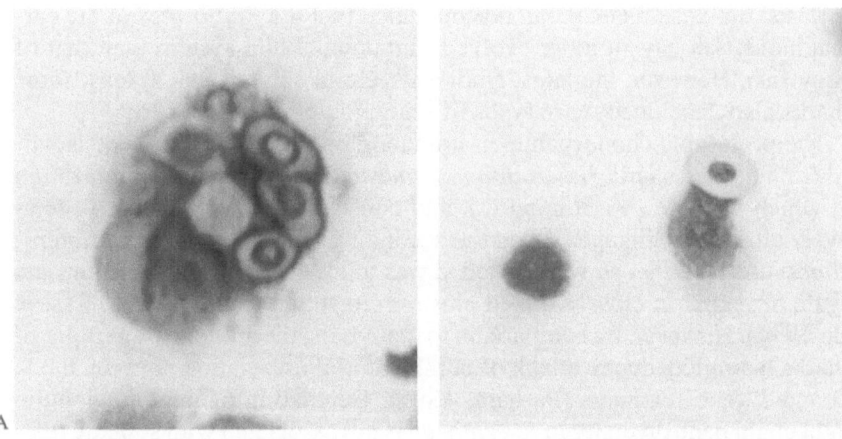

A   B

Figure 7. Phagocytosis of target cells by neutrophils. A) Ingestion of *Lutjanus griseus* erythrocytes by a neutrophil after 4 h of incubation. At least four ovoid erythrocytes can be distinguished within the phagocytic cell. B) Binding of a monocyte to a *L. griseus* target after 18 h of incubation. Magnification × 1000. (From McKinney and Flajnik, 1977, with permission.)

and Flajnik, 1997). These Fc receptors bind both shark 7S and 19S IgM (see Fig. 8) (McKinney and Flajnik, 1997).

Better evaluation of the cytotoxic potential of shark peripheral blood leukocytes has been made possible by findings of better conditions for their cultivation *in vitro* (McKinney, 1992). Both spontaneous and antibody-mediated cytotoxicity have been described in sharks (Pettey and McKinney, 1981, 1983). The spontaneous cytotoxicity is carried out by macrophages. When macrophages were studied, the cytotoxicity was found measurable after 2 h and was complete after 6 h. However, no phagocytosis was observed. On the other hand, shark neutrophils were involved in antibody-mediated cell death via phagocytosis (McKinney and Flajnik, 1997). Down regulation of macrophage-mediated cytotoxicity was demonstrated to be temperature-sensitive and occurred when animals were maintained at temperatures greater than 26°C (Haynes and McKinney, 1988, 1991).

Nurse shark leukocytes migrate in response to a whole variety of chemotactic stimuli (Obenauf and Smith, 1985).

Specific cellular immunity
Tissue rejection has been studied in the horn shark, *H. francisci*. The rejection of allografts was chronic-rejection type (Borysenko and Hildemann, 1970).

The response to mitogens was studied in details is stingray (Olson, 1967) and in nurse shark (Lopez et al., 1974). When the peripheral blood cells were separated on a gradient, the upper band cells responded to Con A, whereas the lower band cells responded to PHA and Con A. An interesting

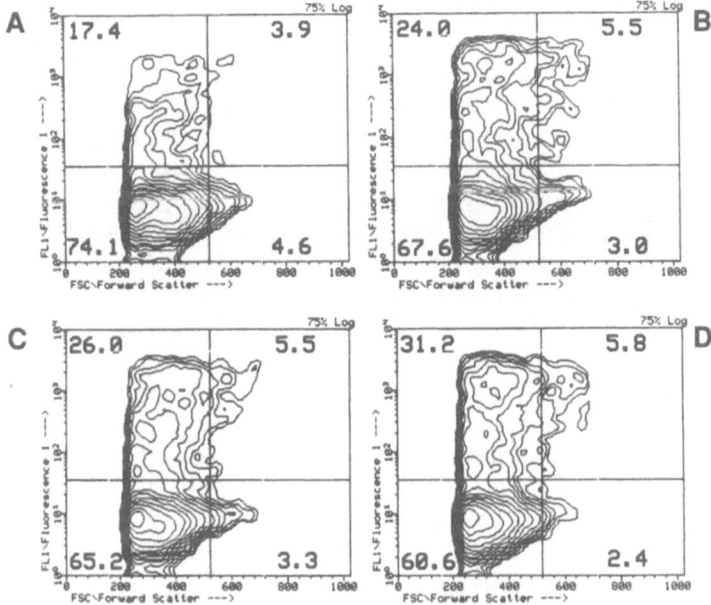

Figure 8. Fcμ receptors bind both 7S and 19S shark IgM. B lymphocytes stained with CB11/16 are shown in panel A. Leukocytes were preincubated with shark 7S IgM (B), 19S IgM (C), or shark plasma (D) before CB11/16 staining. The horizontal line represents the upper limits of the control without CB11/16. The vertical line was arbitrarily drawn to mark the prominent change in fluorescence of the granulocytic population. (From McKinney and Flajnik, 1997, with permission.)

point has been made when the unseparated cells were used. These cells responded only to Con A, suggesting the blocking effect of Con A-responding cells on PHA-responding cells.

Kandil et al. isolated a low molecular mass polypeptide (LMP) complementary DNA clones from the nurse shark (Kandil et al., 1996). This species has two LMP7 genes. In contrast to these findings, hagfishes and lampreys have no LMP genes. Pairwise amino acid sequence comparison and phylogenetic tree analysis showed that the LMP7 gene might have emerged after the appearance of the jawless fish. It is of interest to note that the cartilaginous fish have both MHC class I and II genes (Hashimoto et al., 1992, Kasahara et al., 1993), whereas no MHC genes have been identified in the jawless fish. LMP7 is an MHC-encoded subunit, which is incorporated into the proteasomal complex, involved in MHC class I antigen presentation.

Isolation of a new member of the immunoglobulin superfamily from the nurse shark, which contains one variable and five constant domains and is found as a dimer in serum, suggests that rearranging loci distinct from immunoglobulin and TCR have arisen during evolution (Greenberg et al., 1995).

Conclusions

Chondrichthyes are the first vertebrates exhibiting true adaptive immune response. The immune profile of cartilaginous fish has not been studied in full detail but it is supposed to be comparable to bony fish and in many aspects to tetrapods and even mammals. Both the elasmobranchials and holocephalans possess a thymus in which T lymphocytes mature and are released throughout the body. Recently, it has been shown that the diversity of shark T receptors arises from the same genetic background that gives rise to the diversity of immunoglobulins (Litman et al., 1993). The spleen with red and white pulp fulfils the similar role in lymphohemopoiesis and represnets a source of B lymphocytes producing antibodies in response to antigen stimulation. On the other hand, there is a lack of affinity maturation for shark IgM antibody during the immune response, graft rejection is of a chronic nature, and no true effector functions of T lymphocytes have been described. Antibodies of all gnathostomes ranging from chondrichthyans to endotherms are constructed in the same general fashion of L and H chains in which the tremendous diversity of V region specificities for particular antigenic variants are expressed (Marchalonis, 1977). The sharks and skates have four different immunoglobulin classes, suprisingly more than bony fish, of which only one is shared with mammals. Despite the fact that the cartilaginous fish possess all fundamental anatomical prerequisites for T and B cooperation, the direct proof of the existence of clearly distinct T and B subpopulations is still lacking.

Generally, the cartilaginous fish represent great progress in the evolution of tetrapod adaptive immunity. For the first time in evolution, a clear dichtomy between the central and peripheral lymphoid organs evolved and immunocytes became specialized for precise immune functions. In this grouping of animals, the first mechanisms promoting the diversity of defense together with receptor molecules from immunoglobulin superfamily were developed.

References

Anderson, M.K., Shamblott, M.J., Litman, R.T. and Litman, G.W. (1995) Generation of immunoglobulin light chain gene diversity in *Raja erinacea* is not associated with somatic rearrangement, an exception to a central paradigm of B cell immunity. *J. Exp. Med.* 182: 109–119.

Barre, S., Greenberg, A.S., Flajnik, M.F. and Chothia, C. (1994) Structural conservation of hypervariable regions in immunoglobulins evolution. *Nat. Struct. Biol.* 1:915–920.

Beard, J. (1902) The origin and histogenesis of the thymus in *Raja batis. Zool Jahrb. (Anat. Ontog.)* 17:403–480.

Bernstein, R.M., Schluter, S.F., Shen, S.X. and Marchalonis, J.J. (1996) A new high molecular weight immunoglobulin class from the carcharhine shark: implications for the properties of the primordial immunoglobulin. *Proc. Natl. Acad. Sci. U.S.A.* 93:3289–3293.

Borysenko, M. and Hildemann, W.H. (1970) Reaction to skin allografts in the horn shark. *Transplantation* 10:545–551.

Carroll, R.L. (1988) *Vertebrate Paleontology and Evolution.* W.H. Freeman, New York.

Chandler, A.C. (1911) On a lymphoid structure lying over the myelencephalon of *Lepisosteus. Univ. Calif. Publ. Zool.* 9:85–104.

Chiaje, S. (1840) Anatom disamnine sulle Torpedini. *Atti del Real Instit. d'Incorregiamento Scienze Naturalli di Napoli,* T6:325.

Chibá, A., Torroba, M., Honma, M. and Zapata, A.G. (1989) Structure of the thymic non-lymphoid cells in embryonic and adult elasmobranchs, *7th International Congress Immunol.,* Berlin, Abstract 30–4.

Clem, L.W. and Small, P.A. (1967) Phylogeny of immunoglobulin structure and function. I. Immunoglobulins of the lemon shark. *J. Immunol.* 99:1226–1235.

Colbert, E.H. (1980) *Evolution of the Vertebrates.* John Wiley, New York.

Day, N.K.B., Gewurz, H., Johannsen, R., Finstad, J. and Good, R.A. (1970) Complement and complement-like activity in lower vertebrates and invertebrates. *J. Exp. Med.* 132: 941–950.

Dempster, W.T. (1930) The morphology of the amphibian endolymphatic organ. *J. Morphol.* 50: 71–126.

Drzewina, A. (1905) Contribution d l'etude du tissue lymphoide des Ichthyopsides, *Archs. Zool. Exp. Gen.* 4:145–338.

Ellis, A.E. and Parkhouse, R.E.M. (1975) Surface immunoglobulins on the lymphocytes of the skate, *Raja naevus, Eur. J. Immunol.* 5:726–728.

Fänge, R. (1977) Size relations of lymphomyeloid organs in some cartilagenous fish. *Acta Zool. Stockh.* 58:125–128.

Fänge, R. (1982) A comparative study of lymphoid tissues in fish. *Dev. Comp. Immunol.* Suppl. 2:22–33.

Fänge, R. (1984) Lymphomyeloid tissues in fishes. *Videns. Meddr. Dansk. Natuhr. Foren.* 145: 143–162.

Fänge, R. (1987) Lymphomyeloid system and blood cell morphology in elasmobranchs. *Arch. Biol. Bruxelles* 98:187–208.

Fänge, R. and Johansson-Sjöbeck, M.L. (1975) The effect of splenectomy on the haematology and on the activity of α-6-aminolevulinic acid dehydratase (ALA-D) in haemopoietic tissues of the dogfish, *Scyliorhinus canicula* (Elasmobranchii). *Comp. Biochem. Physiol. [A]* 52:577–580.

Fänge, R. and Mattisson, A. (1981) The lymphomyeloid (hemopoietic) system of the Atlantic nurse shark. *Biol. Bull.* 160:240–249.

Fänge, R. and Sundell, G. (1969) Lymphomyeloid tissue, blood cells and plasma proteins in *Chimaera monstrosa* (Pisces, Holocephali). *Acta Zool. Stockh.* 50:155–168.

Ferren, F.A. (1967) Role of the spleen in the immune response of teleosts and elasmobranch. *J. Florida Med. Assoc.* 54:434–437.

Fujita, T. (1963) Über das Zwischenhirn-Hypophysensystem von *Chimaera monstrosa. Zeitschr. Zellforsch.* 60:147–167.

Gigli, D.M. and Austen, K.F. (1971) Phylogeny and function of the complement system. *Ann. Rev. Microbiol.* 25:309–332.

Good, R.A., Findstad, J., Pollara, B. and Gabrielsen, A.E. (1966) Morphologic studies on the evolution of the lymphoid tissues among the lower vertebrates. *In:* R.T. Smith, P.A. Miescher and R.A. Good (eds): *Phylogeny of Immunity,* University Florida Press, Gainesville, pp 149–167.

Goodman, J.W., Klaus, G.G., Nitecki, D.E. and Wang, A.C. (1970) Pyrrolidonecarbocylic acid at the N-terminal position of polypeptide chains from leopard shark immunoglobulins. *J. Immunol.* 104:260–262.

Greenberg, A.S., Avila, D., Hughes, M., Hughes, A., McKinney, E.C. and Flajnik, M.F. (1995) A new antigen receptor gene family that undergoes rearrangement and extensive somatic diversification in sharks. *Nature* 374:168–173.

Greenberg, A.S., Hughes, A.L., Avila, D., McKinney, E.C. and Flajnik, M.F. (1996) A novel „chimeric" antibody class in cartilaginous fish: IgM may not be the primordial immuno-globulin. *Eur. J. Immunol.* 26:1123–1129.

Hart, S., Wrathmell, A.B. and Harris, J.E. (1986) Ontogeny of gut-associated lymphoid tissue (GALT) in the dogfish *Scyliorhinus canicula* L. *Vet. Immunopathol.* 12:107–116.

Hart, S., Wrathmell, A.B., Harris, J.E. and Grayson, T.H. (1988) Gut immunology in fish: A review. *Dev. Comp. Immunol.* 12:453–480.

Hashimoto, K., Nakanishi, T. and Kurosawa, Y. (1992) identification of a shark sequence resembling the major histocompatibility complex class I $\alpha3$ domain. *Proc. Natl. Acad. Sci. U.S.A.* 89: 2209–2212.

Haynes, L. and McKinney, E.C. (1988) Fc receptor for shark immunoglobulin. *Dev. Comp. Immunol.* 12: 561–571.

Haynes, L. and McKinney, E.C. (1991) Shark spontanous cytotoxicity. *Dev. Comp. Immunol.* 15: 123–134.

Holmgren, N. (1942) Studies on the head of fishes. An embryological, morphological and phylogenetical study. *Acta Zool Stockh.* 23: 129–261.

Hyder, S.L., Cayer, M.L. and Pettey, C.L. (1983) Cell types in peripheral blood of the nurse shark: an approach to structure and function. *Tissue Cell* 15: 437–455.

Jacobshagen, E. (1915) Zur Morphologie des Spiraldarms. *Anat. Anz.* 498: 220–235.

Jarvik, E. (1980) *Basic Structure and Evolution of Vertebrates, Vols. 1 and 2.* Academic Press, New York.

Jensen, J.A., Festa, E., Smith, D.S. and Cayer, M. (1981) The complement system of the nurse shark: hemolytic and comparative characteristics. *Science* 214: 566–569.

Kandil, E., Namikawa, C., Nonaka, M., Greenberg, A.S., Flajnik, M.F., Ishibashi, T. and Kasahara, M. (1996) Isolation of low molecular mass polypeptides complementary DNA clones from primitive vertebrates. Implications for the origin of MHC class I-restricted antigen presentation. *J. Immunol.* 156: 4245–4253.

Kappers, J.A. (1958) Structural and functional changes in the telencephalic choroid plexus during human ontogenesis. *In:* G.E.W. Wolstenholme and C.M. O'Connor (eds): *The Cerebrospinal Fluid,* Churchill, London, pp 3–31.

Kasahara, M., McKinney, E.C., Flajnik, M.F. and Ishibashi, T. (1993) Evolutionary origin of the major histocompatibility complex: polymorphism of class II a chain genes in the cartialginous fish. *Eur. J. Immunol.* 23: 2160–2165.

Kobayashi, K., Tomonaga, S. and Kajii, T. (1984) A second class of immunoglobulin other than IgM present in the serum of cartilaginous fish, the skate, *Raja kenojei*: isolation and characterization. *Mol. Immunol.* 21: 397–404.

Kobayashi, K., Tomonaga, S., Teshima, K. and Kajii, T. (1985) Ontogenetic studies on the appearance of two class of immunoglobulin-forming cells in the spleen of the Aleutian skate, *Bethyraja aleutica*, a cartilaginous fish. *Eur. J. Immunol.* 15: 952–956.

Kobayashi, K., Tomonaga, S. and Tanaka, S. (1992) Identification of a second immunoglobulin in the most primitive shark, the frill shark, *Chlamydoselachus anguineus*. *Dev. Comp. Immunol.* 16: 295–299.

Lacy, E.R. and Reale, E. (1985) The elasmobranch kidney. I. Gross anatomy and general distribution of the nephron. *Anat. Embryol.* 173: 23–34.

Legler, D.W. and Evans, E.E. (1967) Comparative immunology: hemolytic complement in elasmobranchs. *Proc. Soc. Exp. Biol. Med.* 124: 30–39.

Leslie, G.A. and Clem, L.W. (1970) Reactivity of normal shark immunoglobulins with nitrophenyl ligands. *J. Immunol.* 105: 1547–1552.

Leydig, F. (1851) Zur Anatomie und Histologie der *Chimaera monstrosa*. *Müllers Arch. Anat. Physiol.* 10: 241–272.

Leydig, F. (1857) Von Blut und der Lymphe der Wirbeltiere. *In: Lehrbuch der Histologie des Menschen und der Tiere,* Meidinger Sohn, Frankfurt am Main, pp 448–450.

Litman, G.W., Rast, J.P., Shamblott, M.J., Haire, R.N., Hulst, M., Roess, W., Litman, R.T., Hinds-Frey, K.R., Zilch, A. and Amemiya, C.T. (1993) Phylogenetic diversification of immunoglobulin genes and the antibody repertoire. *Mol. Biol. Evol.* 10: 60–72.

Lloyd-Evans, P. (1993) Development of the lymphomyeloid system in the dogfish, *Scyliorhinus canicula*. *Dev. Comp. Immunol.* 17: 501–514.

Lopez, D.M., Sigel, M.M. and Lee, J.C. (1974) Phylogenetic studies on T cells, I. Lymphocytes of the shark with differentiated response to phytohemagglutinin and Concanavalin A. *Cell Immunol.* 10: 287–293.

Makela, O. and Litman, G.W. (1986) Lack of heterogeneity in anti-hapten antibodies of a phylogenetically primitive shark. *Nature* 287: 639–640.

Marchalonis, J.J. (1977) *Immunity in Evolution.* Harvard Press, Cambridge.

Marchalonis, J.J. and Edelman, G.M. (1965) Phylogenic origins of antibody structure. I. Multichain structure of immunoglobulins in the smooth dogfish (*Mustelus canis*). *J. Exp. Med.* 122: 601–618.

Marchalonis, J.J. and Edelman, G.M. (1966) Polypeptide chains of immunoglobulins from the smooth dogfish (*Mustelus canis*). *Science* 154:1567–1568.

Marchalonis, J.J., Hohman, V.S., Thomas, C. and Schluter, S.F. (1993) Antibody production in sharks and humans: a role for natural antibodies. *Dev. Comp. Immunol.* 17:41–53.

Matsunaga, T. and Andersson, E. (1994) Evolution of vertebrate antibody genes. *Fish Shelfish Immunol.* 4:419–423.

Matthews, L.H. (1950) Reproduction in the basking shark *Cetorhinus maximus. Phil. Trans. R. Soc. London B* 234:247–315.

Mattisson, A. and Fänge, R. (1986) The cellular structure of lymphomyeloid tissues in *Chimaera monstrosa* (Pisces, Holocephali). *Biol. Bull.* 171:660–671.

Mattisson, A., Fänge, R. and Zapata, A. (1990) Histology and ultrastructure of the cranial lymphohaemopoietic tissue in *Chimaera monstrosa* (Pisces, Holocephali). *Acta Zool. Stockh.* 71:97–106.

Maximov, A. (1910) Untersuchungen über Blut und Bindegewebe. III. Die embryonale Histogenese des Knochenmarks der Saugetiere. *Arch. Mikrosk. Anat.* 76: S1–113.

Maximov, A. (1923) Untersuchungen über Blut und Bindegewebe. X. Über die Blutbildung bei den Selachiern in erwachsenen und embryonalen Zuständen. *Arch. Mikrosk. Anat.* 97: 623–717.

McKinney, E.C. (1992) Proliferation of shark leukocytes. *In Vitro Cell. Dev. Biol.* 28A: 303–305.

McKinney, E.C. and Flajnik, M.F. (1997) IgM-mediated opsonization and cytotoxicity in the shark. *J. Leukocyte Biol.* 61:141–146.

Moore, K.S., Wehrli, S., Roder, H., Rogers, H., Ogers, M., Forrest, J.N., McCrimmon, D. and Zasloff, M. (1993) Squalamine: an aminosterol antibioticfrom the shark. *Proc. Natl. Acad. Sci. U.S.A.* 90:1354–1358.

Morrow, W.J.W. (1978) The immune response of the dogfish *Scyliorhinus canicula* L. Ph.D. Thesis, Plymouth Polytechnic UK.

Navarro, R. (1987) Ontogenia de los organos linfoides de *Scliorhinus canicula*. Estudio ultraestructural. Master Thesis, Universidad Complutense, Madrid.

Nielsen, M.H., Jensen, J., Braendstru, O. and Werdelin, O. (1974) Macrophage-lymphocyte clusters in the immune response to soluble protein antigens *in vitro*. II. Ultrastructure of clusters formed during the early response. *J. Exp. Med.* 140:1260–1272.

Obenauf, S.D. and Smith, S.H. (1985) Chemotaxis of nurse shark leukocytes. *Dev. Comp. Immunol.* 9:221–230.

Olson, G.B. (1967) Nonspecific and specific lymphoid blastogenesis in leukocyte cultures from *Polyodon spathula* and *Dasyatis americana. Fed. Proc.* 26:357–360.

Parham, P. (1995) Evolutionary immunology. A boost to immunity from nurse sharks. *Current Biology* 5:696–699.

Pettey, C.L. and McKinney, E.C. (1981) Mitogen-induced cytotoxicity in the nurse shark. *Dev. Comp. Immunol.* 5:53–64.

Pettey, C.L. and McKinney, E.C. (1983) Temperature and cellular recognition in the shark. *Eur. J. Immunol.* 13:133–138.

Phisalix, C. (1885) Recherches sur l'anatomie et la physiologie de la rate chez les ichthyopsides. *Arch. Zool. Gen. Exp.* 3:369–464.

Petersen, H. (1908) Beiträge zur Kenntnis des Baues und der Entwicklung des Selachierdarmes. *Jena Zeitschr. Naturwiss.* 43:619–652.

Pilliet, A. (1890) Note sur la distribution du tisse adenoide dans le tube digestif des poissons cartilagineux. *C.R. Soc. Biol. Paris Ser.* 9, 2:593–585.

Policard, A. (1902) Constitution lympho-myeloid du stroma conjonctif, du testicule des jeunes Rajides. *C. R. Acad. Sci. Paris.* 134:297–299.

Pulsdorf, A., Fänge, R. and Morrow, W.J.W. (1982) Cell types and interactions in the spleen of the dogfish *Scyliorhinus canicula* L: an electron microscopic study. *J. Fish Biol.* 21: 649–662.

Pulsdorf, A., Morrow, W.J.W. and Fänge, R. (1984) Structural studies of the thymus of the dogfish, *Scyliorhinus canicula. J. Fish Biol.* 25:353–360.

Pulsford, A. and Zapata, A. (1989) Macrophages and reticulum cells in the spleen of the dogfish, *Scyliorhinus canicula. Acta Zool.* 70:221–227.

Rast, J.P. and Litman, G.W. (1994) T cell receptor gene homologs are present in the most primitive jawed vertebrates *Proc. Natl. Acad. Sci. U.S.A.* 91:9248–9252.

Rast, J.P., Anderson, M.K., Ota, T., Litman, R.T., Margittai, M., Shamblott, M.J. and Litman, G.W. (1994) Immunoglobulin light chain class multiplicity and alternative organizational forms in early vertebrate phylogeny. *Immunogenetics* 40:83–99.

Salkind, J. (1915) Contributions histologuques à la bilogie comparée du thyme. *Arch. Zool. Exp. Gen.* 2:81–332.

Sano, Y. and Imai, C. (1961) Über ein Blutbildenes Organ in der Sella Turcica des Reisensalamanders (*Megalobatrachus japonicus*). *Z. Zellforsch.* 53:471–480.

Scharrer, E. (1897) The histology of the meningeal myeloid tissue in the ganoids *Amia* and *Lepisosteus*. *Anat. Rec.* 88:291–310.

Schluter, S.F., Beischel, C.J., Martin, S.A. and Marchalonis, S.A. (1990) Sequence analysis of homogeneous peptides of shark immunoglobulin light chains by tandem mass spectrometry: correlation with gene sequence and homologies among variable and constant region peptides of sharks and mammals. *Mol. Immunol.* 27:17–23.

Schneider, G. (1897) Über die Niere und die Abdominalporen von *Squatina angelus*. *Anat. Anz.* 13:393–401.

Sigel, M.M. (1974) Primitive immunoglobulins and other proteins with binding functions in the shark. *Ann. NY Acad. Sci.* 234:198–213.

Sigel, M.M., Acton, R.T., Evans, E.E., Russell, W.J., Wells, T.G., Painter, B. and Lucas, A.H. (1968) T2 bacteriophage clearance in the lemon shark, *Proc. Soc. Exp. Biol. Med.* 128: 977–979.

Šíma, P. and Větvička, V. (1990) *Evolution of immune reactions*. CRC Press, Boca Raton.

Šíma, P. and Slípka, J. (1995) The spleen and its coelomic and enteric history. *In:* J. Mestecky, M.W. Russell, S. Jackson, S.M. Michanek, H. Tlaskalová-Hogenová and J. Sterzl (eds): *Advances in Mucosal Immunity, Part A,* Plenum Press, New York, pp 331–334.

Slípka, J. (1979) Phyloembryogenesis of the branchial region. *Pilsen Med. Rep.* 47:5–12.

Slípka, J. and Šíma, P. (1989) Palatine tonsils and its evolutionary ancestors. *Pilsen Med. Rep.* 59:161–165.

Suran, A.A., Tarail, M.H. and Papermaster, B.W. (1967) Immunoglobulins of the leopard shark. I. Isolation and characterization of 17S and 7S immunoglobulin with precipitating activity. *J. Immunol.* 99:679–686.

Stahl, B.J. (1967) Morphology and relatioships of the Holocephali with special reference to the venous system. *Bull. Mus. Comp. Zool.* 135:141–213.

Stanely, H.P. (1963) Urogenital morphology in the chimaeroid fish *Hydrolagus colliei* (Lay and Bennett). *J. Morphol.* 112:99–128.

Tischendorf, F. (1969) Vergleichende makroskopisch-topografische Anatomie der Milz. *In:* Nichtsauger, A. (ed.): *Handbuch der Mikroskopischen Anatomie des Menschen. Blutgefäss- und Lymphgefässapparat innersekretorischer Drüsen. 6. Die Milz.* Springer-Verlag, Berlin, pp 99–100.

Tomonaga, S., Kobayashi, K., Kajii, T. and Awaya, K. (1984) Two populations of immunoglobulin-forming cells in the skate *Raja kenojei*: their distribution and characterization. *Dev. Comp. Immunol.* 8:803–812.

Tomonaga, S., Kobayashi, K., Hagiwara, K., Sasaki, K. and Sezaki, K. (1985) Studies on immunoglobulin and immunoglobulin-forming cells in *Heterodontus japonicus,* a cartilaginous fish. *Dev. Comp. Immunol.* 9:617–626.

Tomonaga, S., Kobayashi, K., Hagiwara, K., Yamaguchi, K. and Awaya, K. (1986) Gut-associated lymphoid tissue in elasmobranchs. *Zool. Sci.* 3:453–458.

Torroba, M., Chiba, A., Vicente, A., Varas, A., Sacedon, R., Jimenez, E., Honma, Y. and Zapata, A.G. (1995) Macrophage-lymphocyte cell clusters in the hypothalamic ventricle of some elasmobranch fish: ultrastructural analysis and possible functional significance. *Anat. Rec.* 242:400–410.

Voss, E.W. and Sigel, M.M. (1971) Distribution of 19S and 7S IgM antibodies during the immune response of nurse shark. *J. Immunol.* 106:1323–1329.

Willmer, E.N. (1970) *Cytology and Evolution*. Academic Press, New York.

Zapata, A. (1980a) Ultrastructure of elasmobranch lymphoid tissue. I. Thymus and spleen. *Dev. Comp. Immunol.* 4:459–472.

Zapata, A. (1980b) Splenic erythropoiesis and thrombopoiesis in elasmobranchs, an ultrastructural study. *Acta Zool. Stockh.* 61:59–64.

Zapata, A. (1981) Ultrastructure of elasmobranch lymphoid tissue. II. Leydig's and epigonal organs. *Dev. Comp. Immunol.* 5:43–52.

Zapata, A. (1983) Phylogeny of the fish immune system. *Bull. Inst. Pasteur,* 81: 165-186.

Zapata, A.G. and Cooper, E.L. (1990) *The Immune System: Comparative Histophysiology.* John Wiley and Sons, Chichester.

Zapata, A.G., Torroba, M., Vicente, A., Varas, A., Sacedon, R. and Jimenez, E. (1995) The relevance of cell microenvironments for the appearance of lympho-haemopoietic tissues in primitive vertebrates. *Histol. Histopathol.* 10:761–778.

Zapata, A., Torroba, M., Sacedon, R., Varas, A. and Vicente, A. (1996) Structure of the lymphoid organs of elasmobranchs. *J. Exp. Zool.* 275:125–143.

Chordates/Vertebrates/Osteichthyes

The osteichthyans, commonly named the bony fish, are characterized by some typical features that set them apart from other vertebrate taxa. The main progressive evolutionary novelties are the ossified endoskeleton, more complex brain, and the air-bladder which is effective in sparing of energy by maintenance of neutral buoyancy.

The members of this huge, specialized class evolved to master many unlikely places such as hot springs or cold circumpolar seawaters with temperatures below the freezing point, and live in marine depths of extremely high hydrostatic pressures or dark subterranean waters. Some groups even began to look for a way to dry land, so they are capable of staying a relatively long time out of water. Because of being under the continuous selective pressure of aqueous environment, the bony fish retained a high degree of conservatism in basic body patterns and remarkable repetition in anatomical details which are shared by all groups independently of their kinship. This developmental rigidity is reflected in the physiology and the immunological profile as well. Only in specialized inhabitants of the diverse water environments mentioned above, can we expect to find a wide variety of physiological functions, including immuno-defense reactions. A basic summary of the immunologically interesting and important nonmorphological features are given in the recent monograph by Šíma and Větvička (1990).

Origins

Within the class *Osteichthyes* only the teleosteans were shown by modern systematics to be of monophyletic origin, whereas the positions of other groups of fishes are still obscure (Patterson and Rosen, 1977; Rosen, 1982; Lauder and Liehm, 1983). Recently, the conclusions based on prevalent morphological characteristics have been confirmed by a row of molecular sequential studies (Normark et al., 1991; Bernardi et al., 1993). Due to enormous diversification and long-lasting evolution of individual fish subtaxa, the investigators engaged in the study of fish immunity must be aware of the evolutionary distances and the phylogenetic interrelationships among the species compared.

The general classification of osteichthyans is surveyed in specialized monographies (Nelson, 1984; Carroll, 1988). The bony fish are divided into two main branches, the ray-finned fish (*Actinopterigii*) and the lobe-finned fish (*Sarcopterygii*). Because of lack of information on the immune behavior of sarcopterygians, we do not deal with them in detail here. The actinopterygian subclass is further divided into two supposed sister groups, the more primitive chondrosteans comprising the lungfishes, sturgeons and paddlefish, and the neopterygians with relatively primitive gars and bowfin, and more advanced teleosteans. Because of a great adaptability, the teleostean fish are unquestionably the most numerous of all vertebrates. For easier orientation within the fish fundamental classification, the kinship among the main osteichthyan subtaxa is shown in Figure 9.

It is supposed that both the chondrichthyans and osteichthyans evolved and radiated independently but simultaneously during Paleozoic. In contrast to Paleozoic elasmobranchian fish, the successful adaptive radiation of osteichthyans continued, so they very quickly dominated the sea environment, more so than all vertebrate groupings. The first fossils of bony fish derive from the Devonian. After partial Cretaceous extinction, the bony fishes and especially the teleosteans have dramatically increased during Tertiary, so they represent about half of all vertebrate species, more than all other classes together, i.e., roughly 20000–25000 species (MacAllister, 1987; May, 1990; Manning, 1994).

The immunocompetent tissues and organs of osteichthyans

Owing to the vast number of fish species, only a very limited sample of modern fish have been studied from the point of structure and function of their lymphoid tissues. Substantial knowledge is available only on fishes of commercial importance living in freshwaters around the continental shelf, whereas the more specialized species such as deep-sea or pelagic fishes are omitted. Despite these limitations and due to the fact that fish maintain conservatively the major structural and functional features, we are more or less able to deduce the general pattern of their immune strategies.

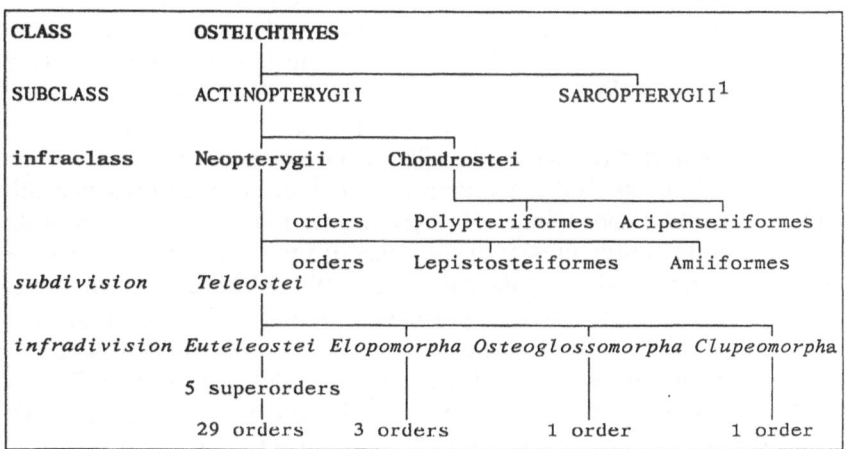

^{1}Not included

Figure 9. Kinships among the main osteichthyan subtaxa.

In fish, like in other advanced vertebrate classes, there are structurally well-developed and functionally specialized internal environments, partly for differentiation of T cells within the thymus, and partly for B cells which develop in secondary lymphoid organs, the spleen, the head kidney, and the trunk kidney, where also trapping, processing, and presentation of antigens by specialized cells take place.

The thymus
Contrary to elasmobranchian multilobulated bilateral thymus, the thymus of bony fish consists of a pair of lobes on each side of the gill cavity. Thymus and pharyngeal epithelium remain in adult teleosteans the permanent integral parts (Ellis, 1977; Grace and Manning, 1980). Whereas the demarcation between cortex and medulla is sometimes unclear in the majority of fish (Ellis, 1977; Zapata, 1981; Tamura and Honma, 1970), in some species well-developed cortex and medulla have been described. There are several modifications, e.g., in *Astyanax* (Hefter, 1952), *Tilapia* (Sailendri and Muthukkarruppan, 1975a, b), *Oncorhynchus* (Chilmonczyk, 1983), *Rutitus* (Zapata, 1981), and in *Sicyastes* (Gorgollon, 1983), where three, and even four layered thymus have been referred. Contrary to ultrastructural arrangement, the thymic tissue of fish seems to be less differentiated than that of tetrapods. Similarly to all vertebrates, the thymus of fish underlies the age-induced histopathologic changes strikingly similar to those that occur in mammals (Sailendri, 1973; Cooper et al., 1983), so that it is difficult to find in an adult fish. Hormonal and/or seasonal influences and stress factors induce the changes of thymic tissue, as well. It is necessary to keep in mind especially the last eventuality; capturing and maintaining the experimental animal under laboratory conditions could induce a rapid involution of the thymus.

As in other vertebrates, the fish thymus is the first organ to become lymphoid in the ontogeny. For instance, in *Cyprinus carpio,* the thymus appears at 2 days post-hatching (at 22°C) and it is colonized by lymphocytes at day 7 (Botham et al., 1980). For the first time in phylogeny, the structures identifiable as Hassall's bodies have been reported in *Polyodon spathula* (Good et al., 1966) and *Sicyases sanguineus* (Gorgollon, 1983), even if other workers are convinced of their presence only in endotherms. Appearance of these mysterious structures is often supposed to be a prerequisite of the onset of immune competence in the ontogeny of mammals (Zapata and Cooper, 1990). There has been proved the significance of the thymus presence in both the transplantation immunity and the humoral response against T-dependent antigens (for review see McCumber et al., 1982; Zapata, 1983; Zapata and Cooper, 1990; Manning, 1994). A minor role is attributed to the thymus as a source of B cells. Some evidence for this is inferred from the presence of plasmacytes and proof of antibody formation in the thymus after immunization.

The spleen

There is little information on the spleen in chondrosteans and primitive neopterygians. In some species, the spleen consisting of red and white pulp, the latter fulfilled by dense aggregations of lymphoid and plasma cells, has been described (Good et al., 1966).

Our knowledge of bony fish spleen, howewer, is mainly based on the structural studies of teleosteans. A large body of data are reviewed in excellent monographs (McCumber et al., 1982; Zapata and Cooper, 1990; Manning, 1994). The spleen is distinctly divided into white and red pulps, even if the splenic tissues are structurally less organized. In some cases the red pulp prevails and may include a whole organ (Grace and Manning, 1980, Secombes and Manning, 1980) or, as in the instance of an icefish, *Chaenocephalus aceratus,* which possesses no red blood cells and it is composed only of lymphocytes and macrophages (Valwig, 1958).

The fish spleen has a key role in erythropoiesis, nevertheless, it is a main site where antibody formation proceeds (Yu et al., 1970). Lymph cells are localized around the splenic arteries often closely associated with reticular and myoid cells (Ellis et al., 1976). The absence of germinal centers typical for all ectotherms is compensated by melano-macrophage centers which are composed of pigment-containing cells, large pyroninophilic cells, and immigrated macrophages and lymphoid cells from the peripheral blood. The macrophage accumulations together with melanin, degraded products of hemoglobin and substances of polyunsaturated fatty acids origin are typical features of the melano-macrophage centers (Agius and Agbede, 1984; Brown and George, 1985). Despite the melano-macrophage centers are mainly engaged in the metabolism of products of erythrocyte breakdown and their renewal during erythropoiesis (Yu et al., 1971; Graf and Schluns, 1979), it is believed that these structures play an immune role similar to germinal centers. The other splenic structures, the ellipsoid sheets, are important in antigen trapping and processing (Secombes and Manning, 1980; Ellis, 1974; Ellis, 1980; Agius, 1981) (see Chapter *Chondrichthyes*). Ellipsoids are found in most teleost fish (for review see Tischendorf, 1969; Zapata and Cooper, 1990; Manning, 1994), and together with melano-macrophage centers, they are clearly engaged in immunoreactivity (Ellis, 1980; Secombes et al., 1982a; Herraez and Zapata, 1986; Van Muiswinkel et al., 1991). The appearance of immunoglobulin-containing cells during immune response in these structures approximates functionally the fish ellipsoids to true germinal centers of endotherms, too (Imagawa et al., 1991). In the course of immune response, a considerable increase of lymphoid cells, proliferation, and newly formed plasma cells in the spleen takes place and these processes are accompanied by intensive vascularization (Sailendri and Muthukkappuran, 1975b; Secombes et al., 1982b). To the contrary, even the spleen is considered to be a main lymphoid organ of the bony fish (Yu et al., 1970); the extirpation of the spleen does not influence substantially the intensity of the immune response, for its immunologic potential is effectively compensated by other lymphoid structures (Ferren, 1967). Similarly, some authors do not consider the spleen, which develops as a last organ during genesis of lymphoid system in fish (Ellis, 1977; Botham and Manning, 1981; Tatner and Manning, 1983), to be important in ontogenic development of immune capacity, for the other lymphoid organs (thymus, head kidney) appear as immunocompetent before the spleen structural maturation (Grace and Manning, 1980; Tatner and Manning, 1983). On the basis of proofs that the splenocytes react to T and B mitogens, the presence of T- and B-like cells in fish spleen has been supposed already in the 1970s (Cuchens et al., 1976; Etlinger et al., 1976).

The gut-associated-lymphoid-tissue (GALT)
As in previous chondrichthyans where the structurally organized GALT appeared for the first time in phylogeny, as in bony fish the digestive system is accompanied by a complex system of lymphoid structures in both the mucosa and the submucosa. The lymphoid cells infiltrating gut epithelia and lamina propria (Pontius and Ambrosius, 1972; Kimura and Kudo, 1975; Davina et al., 1980; Rombout et al., 1986, 1989) may be seen in the form of small clusters but without the histoarchitecture typical for mammals. Such accumulations are found in the spiral valve and in the gut of less advanced representatives of chondrosteans and neopterygians (Good et al., 1966; Fichtelius et al., 1968; Veisel, 1973, 1979; Rowley et al., 1988). The gut epithelial cells function similarly to M cells of mammals (St. Louis-Cormier et al., 1984; Georgopoulou and Vernier, 1986). The antigen presenting macrophages (Rombout and Van den Berg, 1989), antibody-forming cells in lamina propria of fish studied after immunization (Pontius and Ambrosius, 1972), and secretory immunity (Georgopolou and Vernier, 1986) have been described. The function of fish GALT has been intensively studied in recent years for commercial purposes of aquaculture (for review see Manning, 1994).

The immunocompetent tissues and organs of osteichthyans comparable to the bone marrow
No bone marrow exists in the bony fish. Nevertheless, they created an organized hemopoietic tissue analogous to that of more advanced vertebrates which is situated mainly in the pro- and mesonephros, in the spleen, and lesser amounts of that dispersed along the digestive tract.

The kidney
A main body of knowledge on the hemopoiesis in fish comes from studies in teleosteans. The pronephros (head kidney), and the mesonephros and opisthonephros (trunk kidney) are main organs where hemolymphopoiesis takes place. The architecture of these organs is composed of ramified blood

sinusoids, and reticular stroma represent a convenient micromillieu for cytopoietic functions like that of a bone marrow (Ellis and de Sousa, 1974; Zapata, 1979, 1981b). The importance of fish kidney in hemopoiesis is apparent already in early ontogenetic stages when the pronephros becomes the first place to contain hemopoietic capacity (Ellis, 1982) in the developmental stage in which first lymphocytes appear in the thymus (Ellis, 1977; Grace and Manning, 1980; Manning, 1981). During later ontogeny, when the main function of pronephros, excretion, is lost and transferred to the larval mesonephros and even later to adult opisthonephros, the pronephros becomes a hemolymphopoietic organ. The less substantial hemolymphopoietic capacity occurs also in opisthonephros in the intertubular regions. The presence of stem cells (Al-Adhami and Kunz, 1976; Zapata, 1981b) giving rise to all lineages of blood cells including lymphocytes and plasmacytes (Smith et al., 1970; Zapata, 1979) has been demonstrated. Besides the poietic function, both head kidney and trunk kidney are important immunocompetent organs in which the parenchymal stroma formed by ramifying reticular cells having phagocytic capacity is equally suitable for antigen uptake and processing, and where presence of lymphoid cells of thymic origin, macrophages, melano-macrophage centers and active antibody formation has been proved many times (Smith et al., 1967; Chiller et al.,1969; Pontius and Ambrosius, 1972; Ruben et al., 1977; War and Marchalonis, 1977; Anderson et al., 1979; Rijkers et al., 1980a, b; Miller and Clem, 1984; Secombes et al., 1991).

Other structures
Similarly to chondrichthyans, the hemopoietic foci are found in the regions of cranium, medulla oblongata, anterior spinal cord, and heart in the chondrosteans, and representatives of both primitive neopterygian orders (Chandler, 1911; Scharrer, 1944; Good et al., 1966). Lymphocytes and antibody-forming cells have been demonstrated also in skin epidermis and gills (St. Louise-Cormier et al., 1984; Peleteiro and Richards, 1985). The aggregates of lymphoid cells occur in the periportal area of the liver and among the glandular tissue of the pancreas (Andrew, 1959; Gage, 1976). More recently, the pigment-containing macrophages forming melano-macrophage-like structures have been refered in the heart of teleostean fish, the medaka, *Oryzias latipes* (Nakamura et al., 1993).

Cells

B Lymphocytes
Using two different monoclonal antibodies (WCI4 and WCI12) against carp serum immunoglobulins, Koumans-van Diepen et al. demonstrated B cell heretogeneity. Flow cytometry observations showed three subpopulations of B lymphocytes: WCI4$^+$ 12$^-$ cells, WCI4$^-$ 12$^+$ cells and WCI4$^+$ 12$^+$

cells. During ontogeny of the spleen and the pronephros, WCI4$^+$ 12$^-$ cells formed the majority of B lymphocytes at 14 days of age. After that, their percentage declined steadily. On the other hand, the numbers of WCI4$^-$ 12$^+$ cells increased and starting at the age of 13 weeks formed the majority of B lymphocytes (Koumans-van Diepen et al., 1995).

Fish B lymphocytes probably use signal transduction pathway which is analogous to that of mammals. Membrane immunoglobulin on the channel catfish B cells is non-covalently associated with 64 and 70 kDa molecules. Cross-linking of membrane IgM results in phosphorylation of tyrosine residues (Rycyzyn et al., 1996) and in cell proliferation. Similarly to mammalian B lymphocytes, catfish cells exhibited a rapid increase in intracellular calcium levels as result of IgM cross-linking (van Ginkel et al., 1994). The ability to transduce activation signal together with the short cytoplasmic chain of membrane IgM strongly suggest the possibility of IgM association with accessory molecules involved in transduction of the signal. However, this remains hypothetical as no formal demonstration of such accessory molecules in fish B lymphocytes exists.

Studies of fish leukocyte proliferation mediated by protein kinase C done in the red drum, *Scieaenops ocellatus* confirmed that the major route of signal transduction in mammalian cells, phosphatidyl inositol pathway, takes place also in fish. However, protein kinase C isoforms and lower vertebrates' unique membrane ion pumps may participate in cell cycle regulation (Burnett and Schwartz, 1994).

The establishment of a long-term or immortal cell line in mammals almost always involves either immortalization or transformation. The situation in fish is completely different and transient stimulation without any attempts to immortalize the cells results in establishment of cell lines in more than 95% of the fish used (Vallejo et al., 1991a; Lin et al., 1992). Phorbol ester and calcium ionophore in combination are potent mitogens for both T and B cells. Contrary to Con A, these stimulants do not need myeloid accessory cells. Channel catfish B cell lines developed from peripheral blood were characterized in detail. The established and subsequently cloned lines did not require restimulation, feeder cells, or addition of exogenous fragments for more than 12 months. Southern blot analyses of the parental lines showed multiple μ chain gene rearrangements, suggesting a polyclonal origin. The cloned lines revealed mRNA expression for both the membrane and secreted forms of μ chain, similarly both cloned and uncloned lines produced both membrane and cytoplasmic IgM (Miller et al., 1994).

T Lymphocytes
The panel of monoclonal antibodies against fish cells is limited and T cell lineage markers remained elusive. The first antibodies recognizing catfish (CfT1) and carp thymocytes (WCL9) were described recently (Passer et al., 1996; Rombout et al., 1997). Flow cytometric analysis showed that this

antibody reacts with 30–50% of thymocytes and not with cells from blood, pronephros, spleen or intestine. Based on histological observation, the WCL9-positive cells are considered as cortical lymphocytes. These cells might represent early immature T lymphocytes. CfT1 antigen is expressed on thymocytes, a subpopulation of the lymphoid cells in blood and other lymphopoietic organs, but not on erythrocytes, B cells, or macrophages. Stimulation of blood cells by Con A increased the expression of this antigen (Passer et al., 1996). As these results differ from results obtained in carp, these antibodies probably recognize different T cell lineage antigens.

An interesting attempt to analyze T cell development in fish used characterization of genes encoding polypeptides homologous to the TCR molecules. Patula et al. isolated cDNA clones from the rainbow trout that have similar sequences to amphibian, avian and mammalian TCR β chains (Partula et al., 1995). Detailed analysis of three Vβ segments belonging to different families showed some amino acid sequence similarity to the human Vβ20 family. This type of study is interesting from two completely different points of view; one being, of course, the importance of the increased knowledge of phylogeny of TCR, the second (even if rather long-term) being the potential significance in fish farming. Several cDNA segments encoding the TCR has been cloned in the rainbow trout (Partula et al., 1994). These clones encode identical N-terminal-truncated V β region representing limited sequence similarities with several mammalian TCR V β chains.

MHC Antigens

Despite the fact that studies of the allograft rejection have supported the existence of MHC antigens in fish, the MHC molecules have not been isolated so far. First report of MHC encoding genes in fish was made by Hashimoto et al. in the carp (1990). One gene has reasonable homology to MHC class I heavy chains of avian and mammalian species, while the other was homologous to MHC class I β chain. Since this pioneering work, MHC genes have been isolated in zebrafish (Ono et al., 1992; Sultmann et al. 1993, 1994), rainbow trout (Juul-Madsen et al., 1992), Atlantic salmon (Grimholt et al., 1993), striped bass (Walker and McConnell, 1994), guppy (Sato et al., 1995), and channel catfish (Godwin et al., 1997). For a detailed review of isolation and sequencing of genes encoding a chain of MHC class I, β_2-microglobulin and the α and β chain of MHC class II molecules in several fish species see Dixon et al. (1995). The sequences have been used for phylogenetic analysis. A phylogenetic tree constructed using the derived amino acid sequences of the most conserved domain for representatives of teleost and shark MHC genes is shown in Figure 10. Two different cDNA sequences for MHC class II β chains from the channel catfish were identified, sequenced and compared to MHC class II B genes in other

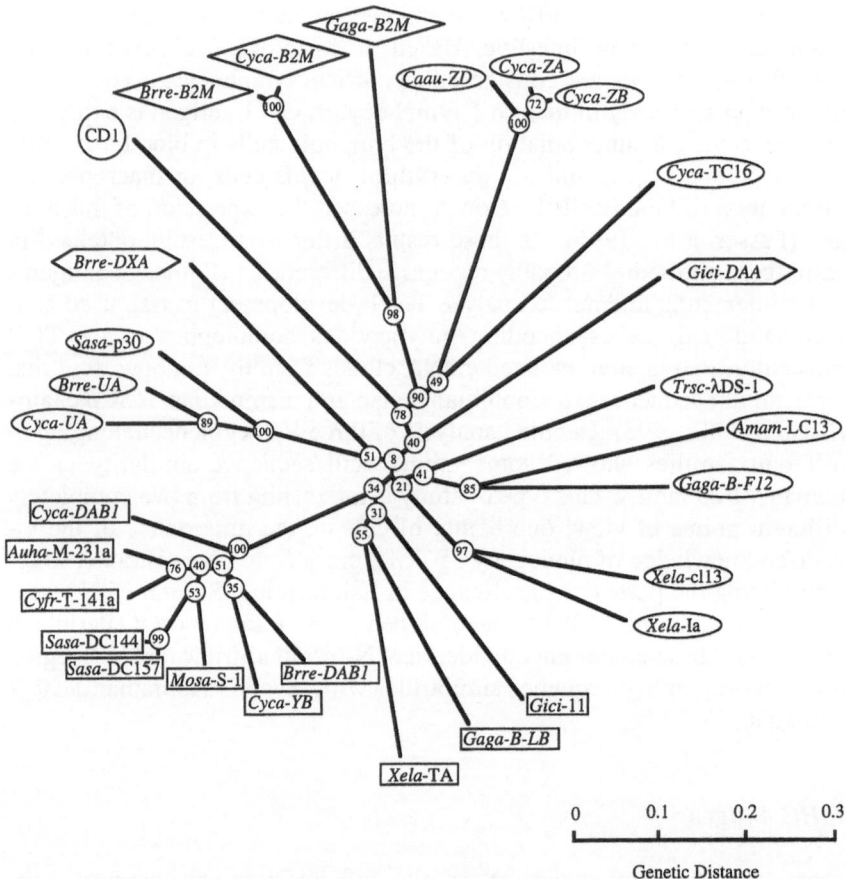

Figure 10. A demonstration of human β_2-microglobulin specific assocation with the surface of teleostean cells. A phylogenetic tree constructed using the derived amino acid sequences of the most conserved domain for representatives of teleost and shark MHC genes (i.e., α_3 domain from class I, α_2 domain from class II, and the β_2-microglobulin domain). This tree was constructed using the computer program Clustal V, which employs the neighbor-joining method. The β_2-microglobulin sequences are indicated by a diamond, the class I α sequences by an ellipse, the class II M sequences by a hexagon, class II β sequences are indicated by a rectangle, and human CD1 is encircled. Bootstrap values for each branching are indicated inside the small circles. (From Dixon et al., 1995, with permission.)

teleost, and showed that they represent alleles of the DAB locus. Southern blot analysis revealed polymorphism in these genes and suggested the presence of two to four different genes (Godwin et al., 1997).

So far, the organization of fish MHC proteins seems to be similar to those of mammals. Using the knowledge of the carp full length MHC class II β chain gene, the MHC class II expression has been studied in different cell types and tissues. From the organs with known immunological role, the highest levels of MHC class II β transcripts was found in the thymus (Rodrigues et al., 1995). The other lymphatic organs including spleen, head kidney, second gut segment and blood had similar levels of transcripts' expression, regardless of their different cellular organization. When the isolated cell populations were used for the same type of experiments, close correlation between sIg$^+$ cells and the high levels of expression were found. However, the highest levels of class II expression were found in thymus leukocytes having low or no surface immunoglobulin expression.

Trans-species polymorphism (sharing of allelic lineages by related species) is a characteristic feature of the major histocompatibility complex. Besides rodents and primates, this feature also exist in teleost fish. The polymorphism is concentrated in the peptide-binding region sites and is maintained by balancing selection. Klein implies that "sharing of this unique MHC feature by both bony fish and mammals suggests that the main function of the MHC (presentation of peptides to T lymphocytes) has not changed during the last 400 million years of its evolution" (Graser and Secombes, 1990).

Purified β_2-microglobulin and its fluorescent conjugates were used for demonstration of the β_2-microglobulin association with the surface of fish cells (Kunich et al., 1994). Employing cell lines from four different species, flow cytometric analysis showed binding of human β_2-microglobulin to the surface of all cell line tested (Fig. 11). A 45 kDa protein co-precipitated with the the β_2-microglobulin after immunoprecipitation with the anti-β_2-microglobulin specific antibody. This protein might be the first teleost MHC molecule isolated so far.

Nonspecific humoral immunity

The non-specific part of the fish immune response involves lectins, various lytic enzymes, C-reactive protein and complement.

The lectins have not been studied in sufficient detail. Based on available literature, natural agglutinins have been detected in representatives from all classes of fish. Their specificity is oriented towards carbohydrate moieties on surface of erythrocytes. For more information, see Síma and Větvička (1990). An interesting lectin having multimeric structure was recently isolated from rainbow trout serum and plasma. Its binding activity is Ca-dependent and can be inhibited by glucose, N-acetyl-glucosamine, mannose, fucose and maltose. When analyzed by SDS-PAGE under non-reducing conditions, the lectin appears as a characteristic ladder of bands, thus its name ladderlectin (Jensen et al., 1997a). It has a multimeric structure composed of variable number of identical polypeptides with MW of 16 kDa. This degree of multimerization is rare and only four similar examples are known. The structural similarities to von Willebrand factor and multi-

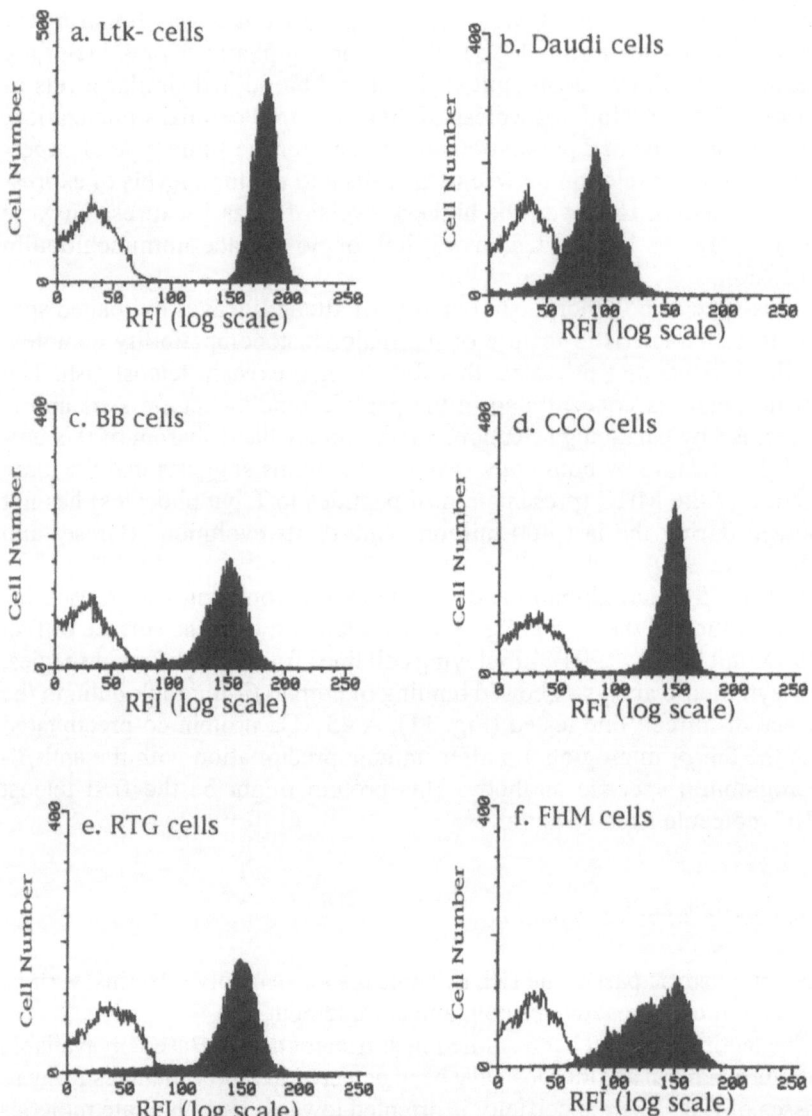

Figure 11. A demonstration of human β_2-microglobulin specific association with the surface of teleostean cell Overlaid fluorescence histograms of live cells incubated with and without β_2-microglobulin-CY5 for 5 h. The x-axis represents relative fluorescence intensities (RFI), cell numbers are plotted on the y-axis. (From Kunich et al., 1994, with permission.)

merin led the authors to the theory that ladderlectin might be involved in blood coagulation. Saha and coauthors studied the presence of antibody dependent hemolysin and opsinin in sera of *Cirrhina mrigala, Clarias batrachus, and Heteropneustes fossilis* and found that these opsonins lyse rabbit erythrocytes (Saha et al., 1993). The heat sensitivity and Ca$^+$ dependency suggest the involvement of fish complement.

Hematopoietic organs of fish contain cystein-rich polypeptides named granulins (Belcourt et al., 1993). These compounds are structurally related to the epithelin family of growth modulatory factors. Granulins have pleiotropic effects on the growth of various fish cells and are probably involved in wound healing and regeneration. Immunohistochemical observations of the granulin-1 localization in the teleost fish *C. carpio* and *Carrasius auratus* showed intensive staining within presumptive macrophage cells in all organs involved in the first line of defense reactions, including skin, gills, gut, heart, spleen and head kidney (Belcourt et al., 1995).

C-reactive protein is a common serum protein found in birds and mammals in response to infection, mostly as result of LPS presence, thus being a classical acute phase reactant. Serum amyloid P component together with C-reactive protein belongs to the family of pentraxins, proteins with characteristic structure of five polypeptides forming a pentagon. Most of our available information was found using mouse and human models. Fish C-reactive proteins were studied in detail in rainbow trout (Winkelhake and Chang, 1982).

Serum amyloid P-component-like molecule was isolated from rainbow trout serum. The NH$_2$-terminal sequence of 16 amino acids has 60% identity with the first residues of human serum amyloid P-component and antibodies against rabbit serum amyloid P-component reacted with this protein (Jensen et al., 1995). Further studies showed approximately 40% amino acid identity of salmonid pentraxin to both mammalian and serum amyloid A and C-reactive protein (Jensen et al., 1997b). Evolutionary analysis suggest the presence of only a single such protein in fish.

The complement system in fish is essentially similar to that of higher vertebrates. Both alternative and classical pathways of complement activation are present, at least the crucial components of the complement system are structurally and functionally similar to their mammalian counterparts. Generally, fish complement is involved in phagocytosis (as opsonin) and in cell lysis (through formation of a membrane attack complex). Saha et al. studied three species belonging to the carp family, *Cirrhina mrigala, Clarias batrachus*, and *Heteropneustes fossilis*, and found that sera not only agglutinate rabbit erythrocytes, but also lyse them. This spontaneous lysis was heat sensitive and Ca-dependent, suggesting involvement of the classical pathway. Tests of CH50 and APCH50 levels showed higher levels of APCH50 (Saha et al., 1993), which is opposite to the situation found in higher vertebrates.

Trout serum contains three functional C3 proteins, C3-1, C3-3, and C3–4, which are the products of at least two different genes. These proteins are composed of two chains with a thioester bond in the α-chain, but differ in their molecular weights, glycosylation, antibody reactivities and amino acid sequences (Sunyer et al., 1996). Subsequent studies were undertaken to find out if the presence of multiple forms of C3 is unique to rainbow trout or if it is a common feature among fish (Sunyer et al., 1997). In the diploid gilthead sea bream, *Sparus aurata,* five different forms of C3 were identified. Figure 12 shows the elution profile, SDS-PAGE, and Western blot analysis of the various C3 and C5 proteins. In addition, a C5-like molecule was characterized. Each of these proteins was forms of an α-chain and a β-chain, but only the V3 isoforms contained a thioester bond in their α-chain (see Fig. 13). Similar to the situation in trout, all these proteins differ in the molecular masses, glycosylation patterns, reactivity with specific antibodies, tryptic peptide maps, and NH_2-terminal sequences in their chains (Fig. 14). There seem to be no similarities in biochemical properties between trout and sea bream C3s, only functional homologies. These findings strongly suggest that these C3 isoforms are produced by several genes. The authors conclude that the isolation of multiple forms of C3 in two fish species which are very distant in evolution suggests that the C3 isoforms generated once from a single C3 gene have remained functional in the genomes of these animals (Sunyer et al., 1997).

With regard to the terminal components of complement, several components such as C5 (Nonaka et al., 1991) or C8 and C9 (Uemura et al., 1996) have been isolated. However, the components of the terminal membrane attack complex (MAC) in fish were isolated only recently (Nakao et al., 1996). Using deoxycholate extraction from rabbit red blood cells lysed by carp serum and subsequent two step chromatography, the authors isolated MAC. On two-dimensional SDS-PAGE of carp MAC, eight bands were found (Fig. 15). Using identifications such as western blotting, *N*-terminal amino acid sequences (Fig. 16), molecular weight under reducing and nonreducing conditions and single-chain structures, carp C5bα, C5bβ, C6, C7, C8α, C8β, C8γ and C9 components were described. Summaries of their molecular weights and their molecular ratios are given in Table 5. Based on these data, the authors conclude that, as with mammals, the cytolytic pathway of fish complement is composed of five terminal components, C5-9 (Nakao et al., 1996). However, at least some of these components (C8α, C8β, C8γ) showed no homology with those of human C8 (Uemura et al., 1996).

Carp factor D is an α-globulin with a MW of 29 kDa. The *N*-terminal amino sequence showed 60% homology to mammalian factor D, but no antigenic or functional compatibility was observed (Yano and Nakao, 1994).

Specific humoral immunity

Fish antibody production has been studied for more than 50 years. Not surprisingly, the early reports were mostly devoted to findings of antibodies in different fish species and to development of optimal techniques for antibo-

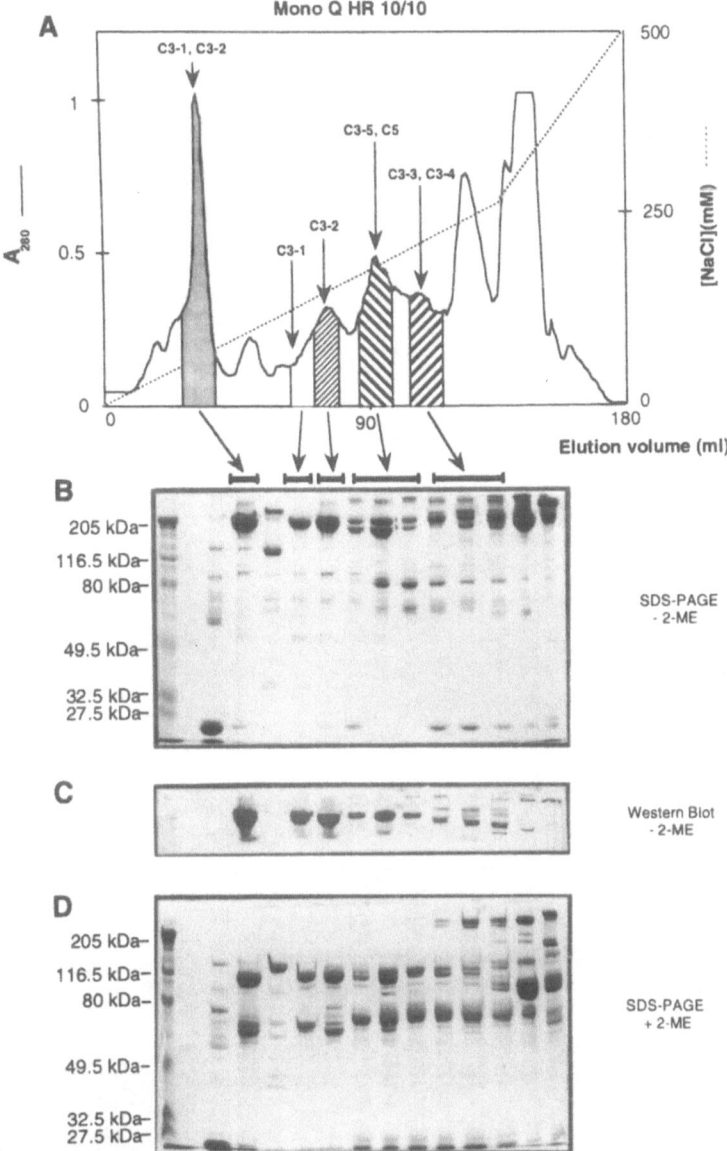

Figure 12. Elution profile, SDS-PAGE, and Western blot analysis of the various C3 and C5 proteins after Mono Q HR 10/10 column chromatography. (A) The 16% PEG fraction of the gilthead sea bream plasma was resuspended in 5 mM sodium phosphate, pH 7.5, containing 5 mM EDTA and applied to a MONO Q HR 10/10 anion exchange column. Bound proteins were eluted with an NaCl gradient 5. The shaded peaks represent the pooled fractions containing the various C3 isoforms and the C5-like protein. (B) SDS-PAGE of the eluted fractions under nonreducing fractions; (C) Western blot analysis of the nonreduced fractions; detection was carried out with an antibody that recognizes trout C3-1. (D) SDS-PAGE of the same fractions under reducing conditions, followed by Coomassie blue staining. (From Sunyer et al., 1997, with permission.)

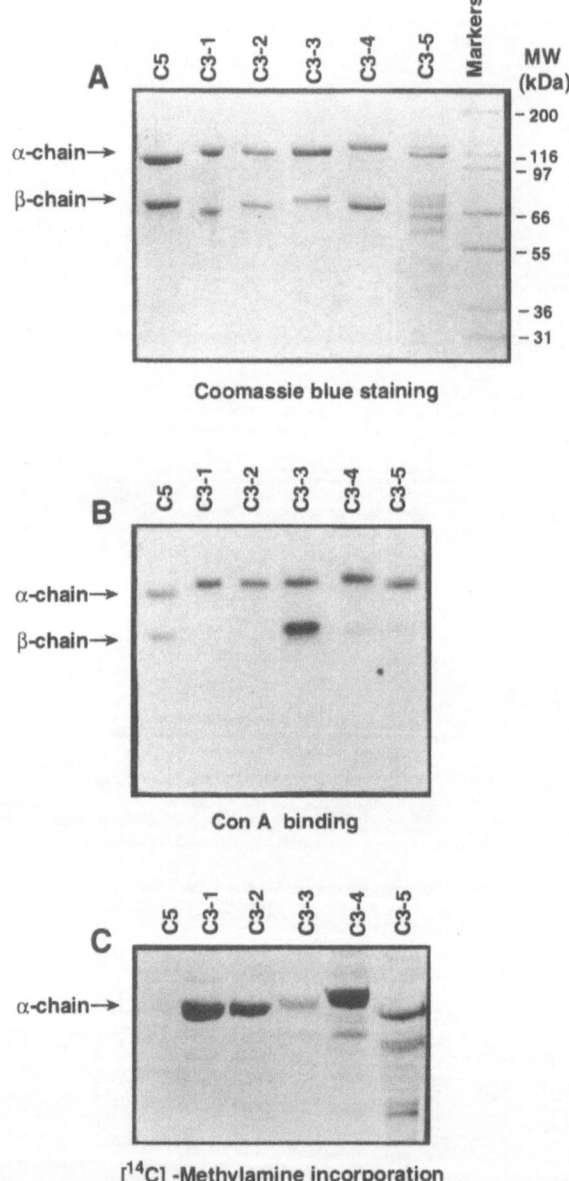

Figure 13. SDS-PAGE, identification of Con A-binding carbohydrates, and analysis of [^{14}C]methylamine incorporation into the various C3 and C5 proteins. (A) Proteins (3 µg) were resolved on 7.5% SDS-PAGE under reducing conditions and stained with Coomassie blue. (B) The same proteins as those in (A) were electroblotted onto a polyvinlylidene difluoride membrane, incubated with ^{125}I-labeled Con A, washed and subjected to autoradiography. (C) After incubation with [^{14}C]methylamine, 10 µg of each protein was electrophoresed, and the gel was treated with EN^{3}HANCER and subjected to autoradiography. (From Sunyer et al., 1997, with permission.)

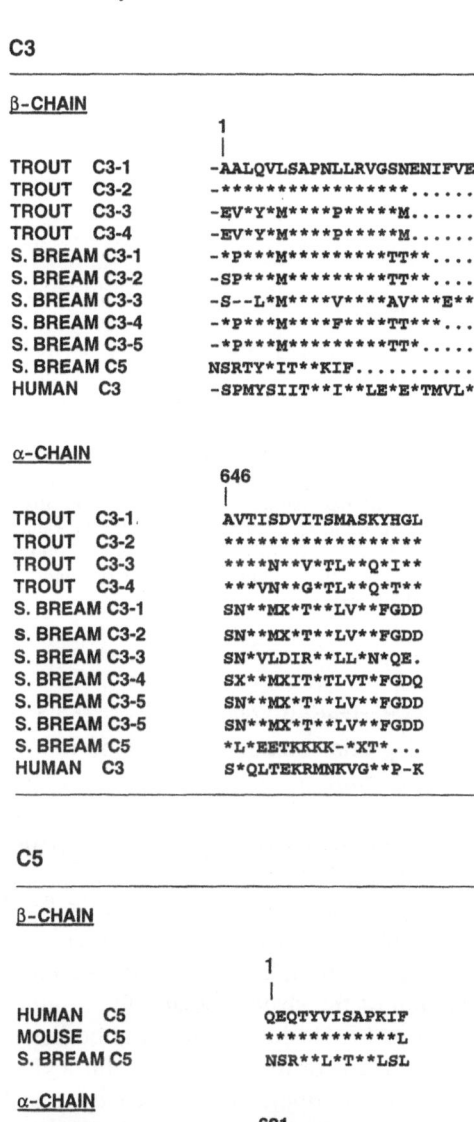

C3

β-CHAIN

```
                             1
                             |
TROUT    C3-1      -AALQVLSAPNLLRVGSNENIFVE
TROUT    C3-2      -****************......
TROUT    C3-3      -EV*Y*M****P*****M......
TROUT    C3-4      -EV*Y*M****P*****M......
S. BREAM C3-1      -*P***M*********TT**....
S. BREAM C3-2      -SP***M*********TT**....
S. BREAM C3-3      -S--L*M***V****AV***E**
S. BREAM C3-4      -*P***M****P****TT***...
S. BREAM C3-5      -*P***M*********TT*.....
S. BREAM C5        NSRTY*IT**KIF...........
HUMAN    C3        -SPMYSIIT**I**LE*E*TMVL*
```

α-CHAIN

```
                             646
                             |
TROUT    C3-1.     AVTISDVITSMASKYHGL
TROUT    C3-2      *****************
TROUT    C3-3      ****N**V*TL**Q*I**
TROUT    C3-4      ***VN**G*TL**Q*T**
S. BREAM C3-1      SN**MX*T**LV**FGDD
S. BREAM C3-2      SN**MX*T**LV**FGDD
S. BREAM C3-3      SN*VLDIR**LL*N*QE.
S. BREAM C3-4      SX**MXIT*TLVT*FGDQ
S. BREAM C3-5      SN**MX*T**LV**FGDD
S. BREAM C3-5      SN**MX*T**LV**FGDD
S. BREAM C5        *L*EETKKKK-*XT*...
HUMAN    C3        S*QLTEKRMNKVG**P-K
```

C5

β-CHAIN

```
                             1
                             |
HUMAN    C5        QEQTYVISAPKIF
MOUSE    C5        ***********L
S. BREAM C5        NSR**L*T**LSL
```

α-CHAIN

```
                             621
                             |
HUMAN    C5        TLQ---KKIEEIAA
MOUSE    C5        N*HLLRQ****Q**
S. BREAM C5        A*TEET**KKAXTY
```

Figure 14. Alignment of the NH$_2$-terminal sequences of the various trout C3, sea bream C3, sea bream C5-like protein, human C3 and C5, and mouse C5. Identical amino acids were indicated by an asterisks; dots indicate regions where amino acid sequence was not determined; dashed indicate gaps introduced to give maximum sequence alignment. (From Sunyer et al., 1997, with permission.)

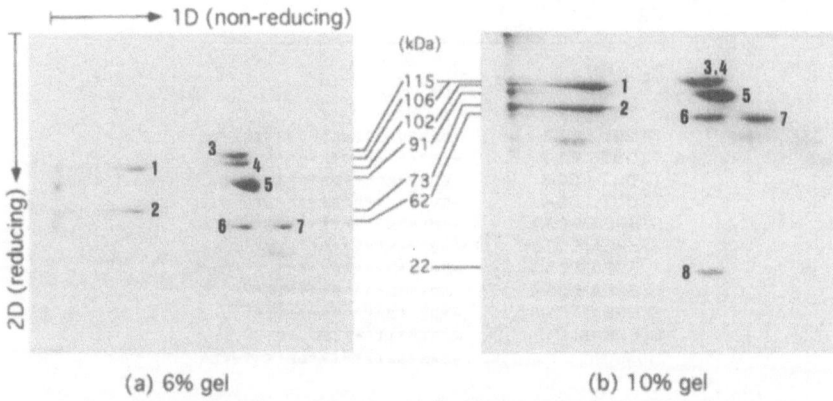

Figure 15. Two-dimensional SDS-PAGE of carp MAC. Purified carp MAC was electrophoresed on a 6% tubular gel under non-reducing conditions and then developed on (a) 6% or (b) 10% slab gel, followed by staining with Coomassie Brilliant Blue. Molecular masses of the eight bands were estimated as follows: Band 1, 102000; Band 2, 73000; Band 3, 115000; Band 4, 106000; Band 5, 91000; Band 6, 62000; Band 7, 62000; Band 8, 22000. 1D, first dimensional run under non-reducing conditions; 2D, second dimensional run under reducing conditions. (From Nakao et al., 1996, with permission.)

dy isolation and characterization. Nevertheless, the purely descriptive report of isolation and partial characterization of fish antibody can be found most recently (Pilstrom and Petersson, 1991; Estevez et al., 1993; Bourmaud et al., 1995). Fish, unlike other vertebrates, synthesize only IgM class antibody, which is tetrameric (in contrast to the pentameric structure known in mammals) without J-chain. The molecular weight is approximately 80 kDa and 27 kDa for the subunits and 850 kDa for the whole molecule. The position and number of interchain disulfide bonds are highly variable between the immunoglobulins of different species (Warr, 1995). A schematic representation of the structures on the immunoglobulins found in fish is given in Figure 17. Besides natural antibodies against common antigens such as DNP haptens or mammalian erythrocytes (Ingram, 1980; Vilain et al., 1984), specific antibodies can be easily raised in fish.

Quite surprisingly, a recent paper published by Wilson and coworkers described a novel complex chimeric Ig heavy chain, partly homologous to the heavy chain of IgG (1997a, b). In addition to alternative secretory or membrane-associated C termini, this molecule also contains a rearranged variable domain, the first C domain of μ, and seven C domains encoded by the delta gene homolog. The authors speculate that IgD is an ancient immunoglobulin molecule that was present in vertebrates ancestral to both mammals and bony fishes. To be able to make any conclusion about this surprising finding, one should wait for an independent confirmation.

Generation of the immunological memory in mammals includes isotype switching and affinity maturation. These processes do not occur in fish, as

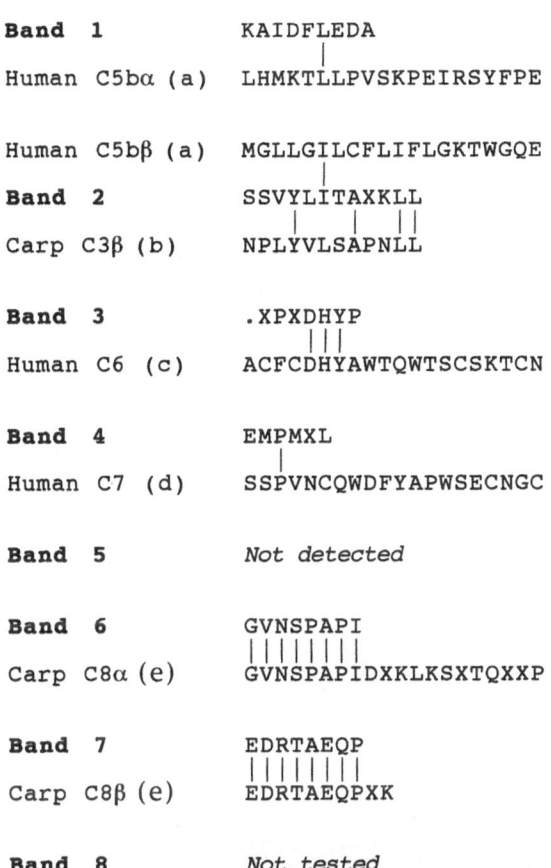

```
Band  1              KAIDFLEDA
                             |
Human  C5bα (a)      LHMKTLLPVSKPEIRSYFPE


Human  C5bβ (a)      MGLLGILCFLIFLGKTWGQE
                             |
Band  2              SSVYLITAXKLL
                             | |  | ||
Carp  C3β (b)        NPLYVLSAPNLL


Band  3              .XPXDHYP
                          |||
Human  C6 (c)        ACFCDHYAWTQWTSCSKTCN


Band  4              EMPMXL
                          |
Human  C7 (d)        SSPVNCQWDFYAPWSECNGC


Band  5              Not detected


Band  6              GVNSPAPI
                          ||||||||
Carp  C8α (e)        GVNSPAPIDXKLKSXTQXXP


Band  7              EDRTAEQP
                          ||||||||
Carp  C8β (e)        EDRTAEQPXK


Band  8              Not tested
```

Figure 16. N-terminal amino acid sequences of the eight protein bands detected on two-dimensional SDS-PAGE. The N-terminal sequences of human C5 to C8 and carp C3β are also shown for comparison. X means unidentified residue and the dot shows the gap inserted for the maximum homology. (From Nakao et al., 1996, with permission.)

Table 5. Identified subunits of carp MAC and their molar ratios

Band #	Identified subunit	Molecular mass	Molar ratio[a]	Molecular mass of human subunit
1	C5bα	102000	0.7 (1)	104000
2	C5bβ	73000	0.9 (1)	75000
3	C6	115000	1	120000
4	C7	106000	1.1 (1)	110000
5	C9	91000	4.2 (4)	71000
6	C8α	62000	1.0 (1)	64000
7	C8β	62000	1.0 (1)	64000
8	C8γ	22000	NT	22000

[a] The molar ratio of subunits to that of C6 was calculated from the molecular mass and protein density on the two-dimentional SDS-PAGE. The values in parentheses show the nearest integer. (From Nakao et al., 1996, with permission.)

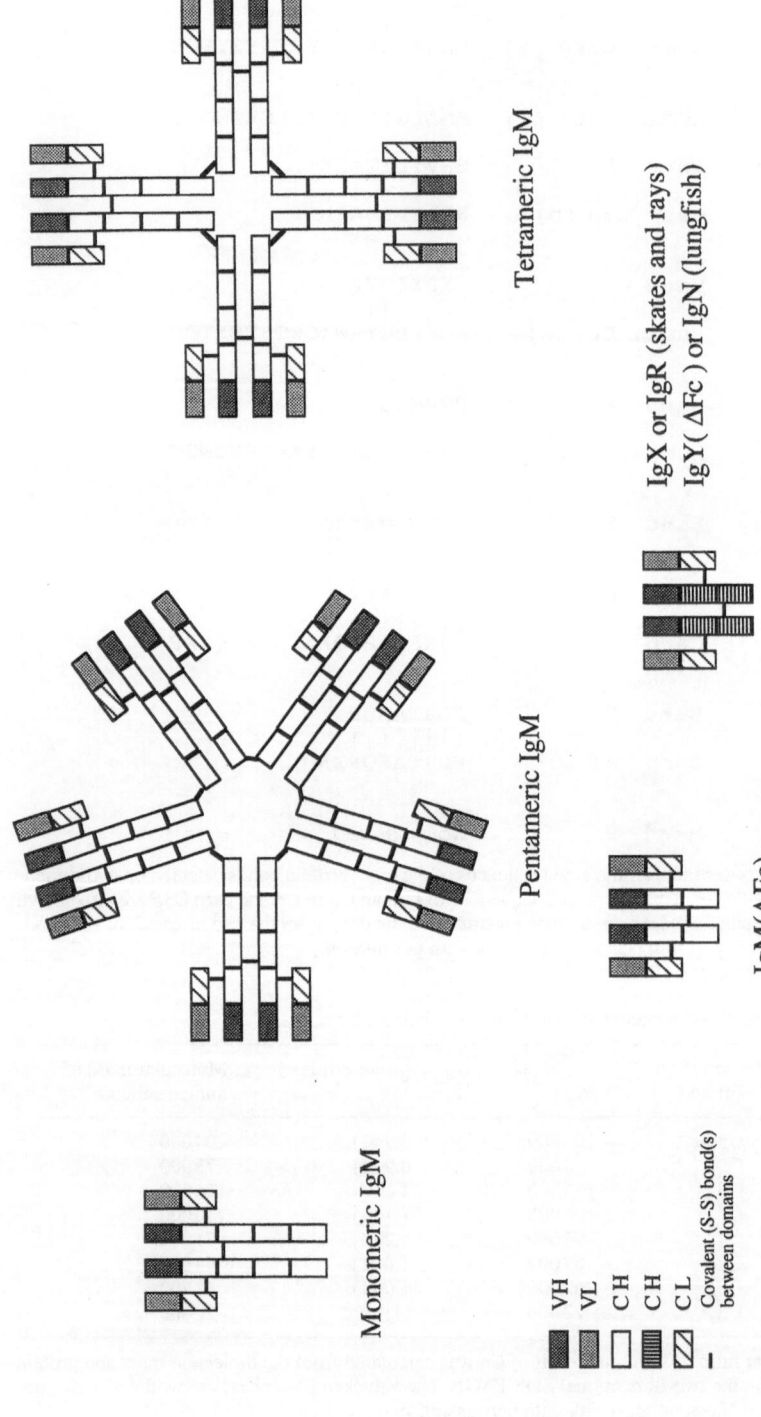

Figure 17. Schematic representation of the structures of the immunoglobulins found in fish. The structures of the IgM (ΔFc) and the antibody of the lungfish are speculative in that they are based on the sizes of the heavy chains rather than on primary structural analyses. The position and number of interchain disulfide bonds are highly variable between the Igs of different species. (From Warr, 1995, with permission.)

they produce only antibody of one isotype. However, immunological memory has been shown in fish (Trump and Hildemann, 1970; Desvaux and Charlemagne, 1981; Tatner, 1986). Enhanced secondary antibody response is due to an expansion of the antigen-sensitive precursor pool without a concomitant increase in clone size (Arkoosh and Kaatari, 1991). To produce an enhanced secondary response, T-dependent antigens required more injections than T-independent antigens. The observations of the possible affinity maturation led to inconclusive results, probably due to the differences between species used in these experiments.

Numerous studies have shown that the heavy chain is rearranged from V, D, J and C genes, and that the organization and structure of these genes are variable. For a more recent review summarizing current knowledge of the structure, organization, and functional expression of immunoglobulin genes in fish, see Warr (1995). Analysis of a complete heavy chain variable region gene isolated from rainbow trout showed that the 98-amino acid long V_H coding region has 50–70% nucleotide sequence homology and 40–60% amino acid sequence homology with variable region genes of various vertebrates (Matsunaga et al., 1990). The genomic organization of H chain gene segments in fish is different from both mammals and sharks. Fish have only a single genomic copy of the $C\mu$ gene (Ghaffari and Lobb, 1989; Ventura-Holman et al., 1994). V_H gene families diverged within the phylogeny of osteichthyes. Eight different catfish V_H gene families have been found, together representing over 120 different V_H gene segments (Ventura-Holman et al., 1996; reviewed in Ghaffari and Lobb, 1997). The organization of the heavy chain locus seems to be a forerunner of the system found in amphibians and mammals, as V_H, D_H, and J_H exons are located in separate regions undergoing somatic reorganization and splicing with a single C_H element. Using a specific probe generated by PCR, *S. salar* cDNA library has been searched and two IgM heavy chains (named C_HA and C_HB) have been cloned and analyzed (Hordvik et al., 1992). The nucleotide and amino acid identities between these two fractions were 98.3% and 96.2 %, respectively. Heavy chain sequences of the IgM molecule were studied in the bowfin, *Amia calva* and the longnose gar, *Lepisosteus osseus*. Each heavy chain showed the conserved four C domain structure. Southern blot analysis using specific V_H and C_H probes support that both fish possess an IgM locus that resembles that of fish, amphibians, and mammals in its organization (Wilson et al., 1995a).

Complementarity determining region CDR3 has been studied in Atlantic cod, *G. morhua*. This region is very heterologenous and gives a major contribution to V_H diversity. Comparison of V_H sequences between 17 fish species showed that different V_H families are often more conserved between species than within any one species (Bentgen et al., 1994).

Both secreted and membrane IgM are encoded by a single gene. The mRNAs encoding these two forms are derived from a single primary transcript by alternate pathways of RNA processing. Warr's group studied the pre-mRNA splicing pathways of the heavy chain transcript of *Ictalurus punctatus* and found that the only detectable splicing pattern used in heavy chain production utilizes the pathway in which the first transmembrane exon is spliced directly to the $C\mu3$ exon and not into a cryptic site within the CH4 exon (Warr et al., 1992). The studies of the structure and patterns of expression of μ genes might be potentially useful in determining the mutual evolutionary relationships between individual species of fish (Wilson et al., 1995). Organization of the IgH loci in fishes is shown in Figure 18. Although no heavy chain switch occurs during the immune response, the affinity of antibodies does increase, suggesting somatic mutations (Ventura-Holman et al., 1994).

Light chain population is heterologous. Two different chains, 22 kDa form designated F, and 26 kDa form designated G were described. cDNA representing germline G transcripts as well as full length cDNA were characterized. The data showed that gene segments are organized in clusters, with V_L, J_L, and C_L segments represented in each cluster. Within each individual cluster, two V_L segments are located upstream of closely linked J_L, and C_L segments (Ghaffari and Lobb, 1993). Subsequent studies have been devoted to the genomic organization of F chain. Based upon these results, different families of V_L segments are associated with closely related F C_L segments. F gene segments are arranged in closely linked clusters with single copies of V_L, J_L, and C_L segments (Ghaffari and Lobb, 1997). Therefore, it seems reasonable to conclude that both L chain classes evolved within a common organizational pattern of clustered genes.

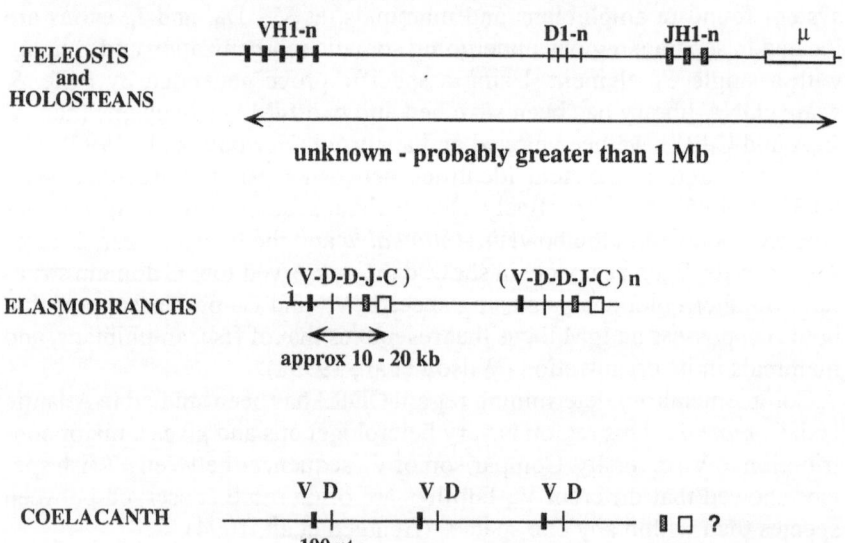

Figure 18. Organization of the IgH loci in fishes. The three main types of IgH locus organization described to date in fishes are shown. (From Warr, 1995, with permission.)

A complete cDNA clone encoding secreted IgM was isolated from the spleen cDNA library of rainbow trout. Amino acid sequence comparison with IgM isolated from other vertebrates revealed that some IgM domains have evolved at a relatively constant rate (Andersson and Matsunaga, 1993).

Mochida and co-workers studied the systemic antibody response to protein antigens. Tilapia produced antibodies with high specificity, but when challenged with multiple protein antigen, it secreted antibodies only to the major components, but not to others, which is quite different from that in mammals (Mochida et al., 1994).

Very few reports have been published about the ontogenetic development of antibody reaction in fish. Nagae et al. used a sensitive ELISA assay to measure changes in IgM level during early development of chum salmon, *Oncorhynhus keta,* and found that the antibody levels remained at concentration below 300 ng/ml until 40 days after hatching and increased dramatically just after emergence (1993).

The effects of temperature on antibody response in fish have been known for a long time (Avtailon, 1981; Clem et al., 1991; Bly and Clem, 1992). Low environmental temperatures decreased antibody production with maximum effects on day 28. Around 50 days after exposure, an adaptation to low temperatures occurred (Morvan et al., 1996). Similar study on the sunshine bass, *Morone saxatilis*, showed significantly lowered antibody response when temperature reached 10°C, but no effect when unusually high at 29°C (Hrubec et al., 1996).

Killie and Jorgensen studied immunoregulation in fish testing different hapten-carrier ratios and combinations of various haptens and carriers in their ability to induce antibody response in salmons. Using a whole panel of T-dependent hapten-carrier antigens such as chicken gamma globulin or hemocyanin coupled with NIP, TNP or FITC they found that the fish responded by high anti-hapten reaction, whereas anti-carrier antibody response was suppressed 87–99% (Killie and Jorgensen, 1994). When NIP and FITC were intramolecularly conjugated to the hemocyanin, NIP induced suppression of the anti-FITC reponse. This suppression was not dependent on haptenation ratios or time of immunization. Even after 134 days the anti-carrier response was negligible in fish immunized with NIP-carrier compared to responses in animals given the carrier alone. Similar suppression was also described in carp (Avtailon and Milgrom, 1976). The fact that the potentially immunodominant determinant located in one part of a molecule may induce intramolecular suppression of other epitopes on the same antigen molecule might be important for immune protection (Killie and Jorgensen, 1994). A subsequent study used both T-dependent and T-independent hapten-carrier antigens with similar results seen with T-dependent antigens, suggesting that the mechanism of intermolecular suppression is T-cell dependent (Killie and Jorgensen, 1995). Using different antigen mixtures differing in amounts of competing antigens showed that the kinetics of suppression was ratio- and dose-dependent. The carrier antigen was able to induce suppression of hapten epitopes, but only when the anti-carrier antibody response was allowed to develop for 14 days prior to hapten-carrier administration. These data show that the intermolecular suppression is a common regulatory phenomena controlling the initial phase of the immune response.

Nonspecific cellular immunity

Like in all animals, phagocytosis is the first line of defense. Bony fish are ectothermic vertebrates, thus macrophages and phagocytes play an even more important role in defense reaction, as lymphocytes are suppressed in cold temperatures (Bly et al., 1990). The study of the effects of cold acclimation and assay temperatures on peritoneal macrophages from carp and goldfish showed that the substratum adherence was undisturbed. On the other hand, dehydrogenases and endocytosis were temperature-sensitive. Thermal acclimation of fish resulted in return to the normal phagocytic values (Plytycz and Jozkowicz, 1994). Besides changes in temperature of the external millieu, phagocytic capacity and superoxide anion production reflects also seasonal changes (Collazos et al., 1995). Rainbow trout peritoneal macrophages actively phagocytose opsonized latex particles and migrate towards fish C5a or FMPL (Zelikoff et al., 1991).

Our knowledge of the fate of indigestible inert material in fish is extremely limited. Encapsulation or nodule formation have been studied almost exclusively in invertebrates. In *Oryzia latipes*, intraperitoneally injected carbon particles were first taken up by phagocytic cells and subsequently concentrated in melano-macrophage centers. The carbon particles persisted in these centers for more than 2 years. On the other hand, subcutaneously administered carbon was gradually eliminated from the skin through trans-epithelial portage by macrophages (Nakamura et al., 1992). Similar results were achieved using an ultrastructural observation of the uptake and handling of *Vibrio salmonicida* in phagocytes of the head kidney from Atlantic salmon (Brattgjerd and Evensen, 1996). Using cardiac endothelial cells of the medaka *O. latipes* and the goldfish *Carassius auratus* as a model of phagocytosis of intraperitoneally injected carbon particles, Nakamura and Shimozawa showed the high level of active phagocytosis by medaka cells, but significantly lower activity by those of the goldfish (Nakamura and Shimozawa, 1994). Phagocytic cells which reside in the heart, have developed cytoplasmis processes extending toward the heart lumen.

The *in vitro* responses of channel catfish neutrophils to *Edwardsiella ictaluri* were tested using phagocytic and bactericidal assays. The functional abilities of fish neutrophils are still unclear compared to their mammalian counterparts, as some reports have stated that these cells are either not or only weakly phagocytic (Ellis, 1981; Ellsaesser et al., 1985). Later tests clearly demonstrated high endocytosis (Waterstrat et al., 1991). In the tench, *Tinca tinca*, granulocytes phagocytize more opsonized *Candida albicans* than nonopsonized yeast, but the killing of yeast was not dependent on opsonization (Pedrera et al., 1992).

These results are in agreement with findings of the phagocytic activity in neutrophils isolated from other fish species (Sakai, 1984; Bodammer, 1986; Suzuki, 1986). However, no killing was observed in bactericidal assays (Waterstrat et al., 1991). The situation probably varies from species to species, as variation among fish in the level of metabolic activation is evident (Dexiang and Ainsworth, 1991). Phagocytosis and intracellular degradation of heterologous proteins was also demonstrated in eosinophilic granulocytes isolated from posterior intestine of the rainbow trout (Ellsaesser and Clem, 1994). More information about phagocytosis in fish can be found in Šíma and Větvička (1990).

A different situation has been found in macrophages. Yeast β-glucan injected *in vivo* stimulates fish macrophages to kill bacteria (Jorgensen et al., 1993). When incubated *in vitro*, rainbow trout and salmon macrophages showed a marked increase in respiratory burst activity, but no increased antibacterial activity (Jorgensen and Robertsen, 1995). The possible explanation might be that to kill *Aeromonas salmonicida*, salmon macrophages require a simultaneous triggering of other bactericidal functions. Besides β-glucan, fish macrophages respond by increased respiratory burst to other substances, such as LPS, tumor necrosis factor or macrophage-activating factor (MAF) (Novoa et al., 1996). MAF is capable of maximal priming after 6 h, but after 24 to 48 h no priming can be observed. Kinetics of the LPS stimulation are different – the increase is gradual up to 48 h. Surprisingly, when MAF and LPS were used simultaneously, the respiratory burst activity was even more enhanced, but the kinetics were similar to those induced by MAF only (Neumann and Belosevic, 1996). Besides reactive oxygen intermediates, stimulated macrophages also produced reactive nitrogen intermediates. This production was biphasic – the cells were able to selectively deactivate one branch without affecting the second one, which might help to minimize host tissue damage.

The production of MAF by leukocytes is temperature dependent (Hardie et al., 1994). On the other hand, macrophages isolated from fish kept at low temperatures and cultured at 6°C responded quite well to MAF addition. Similarly, their respiratory burst activity was higher than that from macrophages cultivated at higher temperature (Hardie et al., 1994). These data confirm the theory that there are reciprocal temperature-dependent activities between macrophages and lymphoid cells.

Zymosan binds to the surface of catfish macrophages and is rapidly phagocytosed. Subsequent studies using FITC-labeled β-glucan and inhibition of phagocytosis demonstrated that zymosan binds to the β-glucan receptor (Ainsworth, 1994). The authors speculated that this receptor might be the fish equivalent of CR3 receptor, CD11b/CD18. When the *in vitro* effects of β-glucan on salmon macrophages were studied, stimulation of superoxide anion formation, increased pinocytosis, increased level of acid phosphatase activity, and increase in cell diameter and spreading were found (Sveinbjorsson and Seljelid, 1994). A tendency to form cell aggregates was described after prolonged incubation with β-glucan (Fig. 19) with the most pronounced effect for the lowest glucan concentrations (Jorgensen and Robertsen, 1995). Intestinal uptake and organ distribution of β-glucan after rectal or intragastrical administration was studied in Atlantic salmon, *Salmo salar.* β-Glucan appeared in blood and tissue 15 min after application. Histological observations revealed that the β-glucan was internalized by epithelial cells in the lower part of the intestine (Sveinbjornsson et al., 1995). Due to the known immunostimulatory effects of β-glucan, these findings suggest the feasibility of use of β-glucan as feed additive in fish culture. Tissue administration of β-glucan was tested in Atlantic cod, *Gadus morhua.* Whole body autoradiography and liquid scintillation showed significant accumulation in heart, spleen, anterior kidney and posterior kidney. Fluorescence microscopy demonstrated presence of glucan-FITC in kidney macrophages, heart endothelial cells and splenic ellipsoid sheath cells (Dalmo et al., 1996). The authors provided no explanation for the rather unexpected accumulation of β-glucan in the heart. The specificity of a β-glucan receptor on salmon macrophages was evaluated by Engstad and Robertsen. Linear 1,3-β-glucans and small oligomers from formolyzed β-glucan were strong inhibitors of phagocytosis of β-glucan (1994), 1,6-β-glucans were not effective. Similarly, endo-β-1,6-glucanase treatment did not influence the phagocytosis. Taken together, these results support the hypothesis that fish macrophages express receptor recognizing 1,3-β-glucan, but not 1,6-β-glucan.

Bovine transforming growth factor $β_1$ (TGF-$β_1$) increased the respiratory burst activity of rainbow trout macrophages. However, when macrophages were either preincubated with some activation stimulus and subsequently treated with TGF-$β_1$ or simultaneously incubated with TGF-$β_1$ and activating substance, a significant inhibition of respiratory burst activity was observed (Jang et al., 1994). The mechanisms are unclear. Based on these results, TGF-$β_1$ can be placed among bioactive molecules showing biological crossreactivity across evolution.

Muramyl dipeptide activates fish phagocytes. Head kidney cell collected from rainbow trout injected with muramyl dipeptide showed high chemotactic activity and elevated phagocytosis. When control macrophages isolated from untreated fish were incubated with supernatants of cell cultures of muramyl dipeptide-treated fish, the cells responded by increase of chemotaxis and phagocytosis, indicating the synthesis and release of an activating factor (Kodama et al., 1993).

Examining the *in vitro* effects of cortisol and estradiol on the inflammatory function of a macrophage cell line derived from goldfish, Wang and

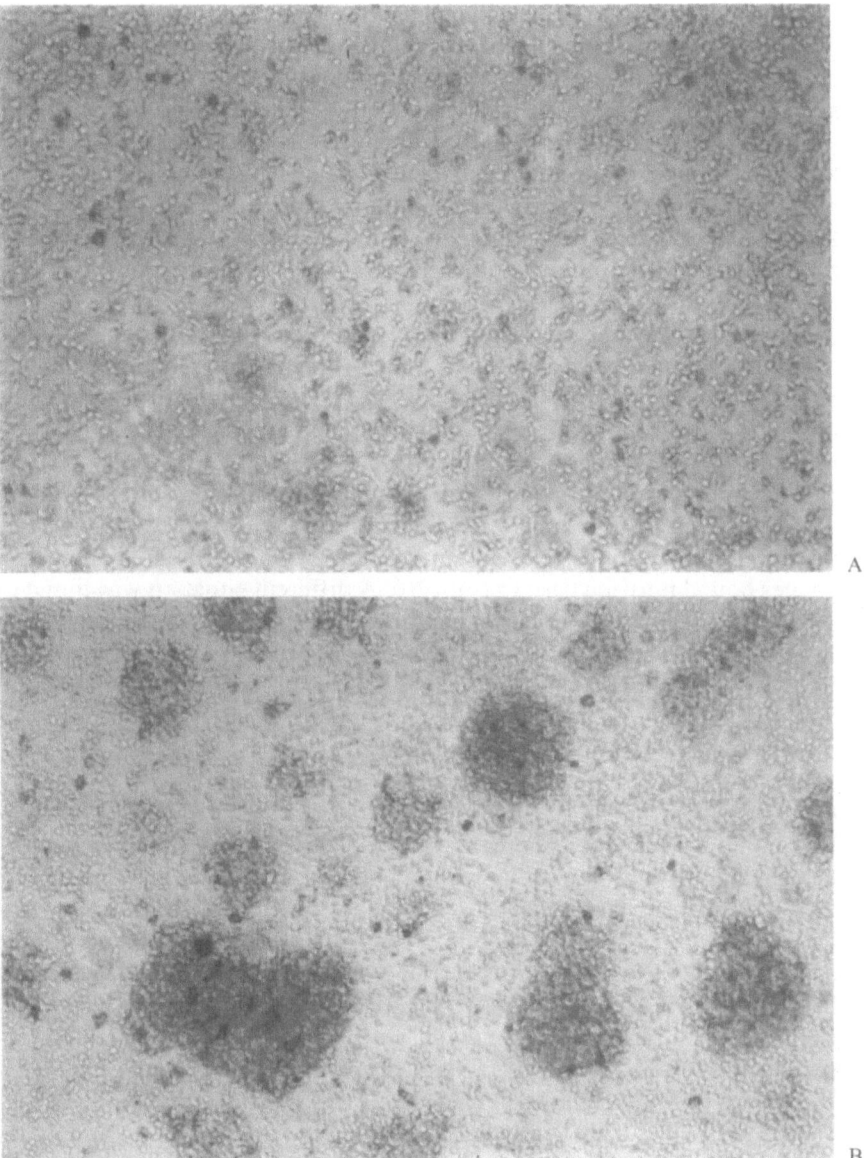

Figure 19. Cell-aggregate formation in macrophages cultures for 72 h in the presence or absence of yeast β-glucan. (A) Control cultures. (B) Cultures of glucan-stimulated cells. Magnification × 85. (From Jorgensen and Robertsen, 1995, with permission.)

Belosevic found a dose-dependent inhibition of chemotaxis and phago-cytosis (1995) and inhibition of the mitogen-induced proliferation of lym-phocytes (1994). On the other hand, susceptibility of goldfish to *Trypano-soma danilewskyi* was increased by estradiol. Estradiol did not influence the nitric oxide production, while cortisol was strongly inhibitory. Respira-tory burst response was not affected (Wang and Belosevic, 1994). The pro-duction of the nitric oxide is further stimulated by macrophage activating factor secreted by mitogen-stimulated goldfish kidney leukocytes. LPS has similar effects and simultaneous administration of both stimulants had syn-ergistic effect (Nagae et al., 1993).

Fish leukocytes respond to chemotactic stimuli such as lipoxin or leuko-triene B_4 by chemokinesis (Sharp et al., 1992). This finding might be potentially very important, as lipoxins are generated in significant quanti-ties by fish macrophages, suggesting their role in communication between immunocompetent cells.

Using immunofluorescence and immunogold staining, the Ig-binding capacity of Ig-positive macrophages has been studied in carp, *Cyprinus carpio*. Thirty minutes after internalization of Ig, most hindgut macro-phages bound purified carp Ig (Fig. 20). A different situation was found in pronephros. A limited number of monocyte-like cells bound and internalize Ig (Fig. 21), but macrophages and neutrophilic granulocytes appeared to be Ig-negative (Koumans-van Diepen et al., 1994). The authors hypothesize that carp intestinal macrophages express putative FcR, while pronephros cells lack this receptor.

Cytokines

In mammalian immune system, interleukins play a pivotal role is intracellu-lar regulations. Our knowledge of the interleukin pathways in fish is still only rudimentary. The presence of IFN-γ has been proposed in trout (Gra-ham and Secombes, 1990) and an IL-2-like factor has been found in carp leukocytes (Grondel and Harmsen, 1984). Similarly, IL-1-like molecule has been shown in salmon (Hamby et al., 1986) and catfish (Ellsaesser and Clem, 1994). Later studies showed that carp granulocytes and macrophages secrete an IL-1-like factor, which induces proliferation of both carp T-cells and mouse IL-1-dependent cell line (Sigel et al., 1986, Verburg – van Keme-nade et al., 1995). This factor can be immunoprecipitated using antibodies against human IL-1. Further analysis revealed that at least two forms of IL-1-like activity exist, a high molecular weight form (70 kDa) active on chan-nel catfish cells, but not on mouse cells, and a low molecular weight form (15 kDa) active on mouse, but not catfish cells. Both forms have α and β determinants as determined using specific antibodies and antibody inhibiti-on assays (Ellsaesser and Clem, 1994). The reasons why fish and mouse dif-fer in response to these two fractions remain to be answered.

Recombinant human tumor necrosis factor (TNF-α) was found to elevate the respiratory burst of trout macrophages, especially in cooperation with

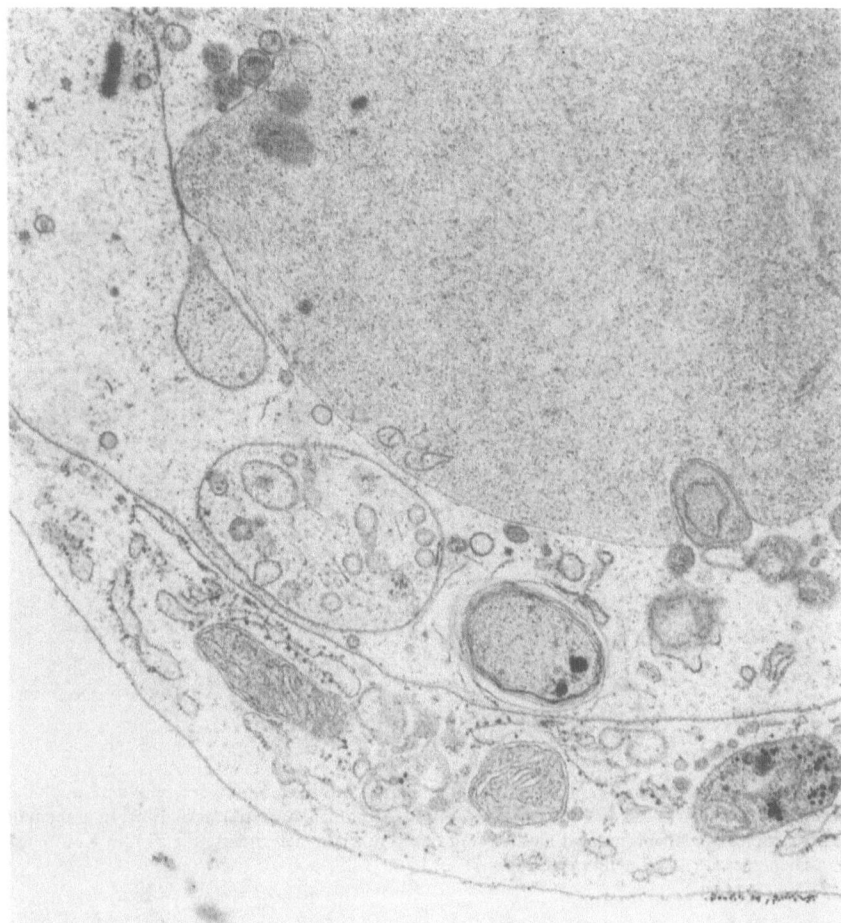

Figure 20. Electron micrograph of a large hindgut macrophage labeled with a monoclonal antibody against Ig (WCI 12). Large gold particles (30 nm) represent internalized Ig within endosomal structures and small gold particles (10 nm) represent newly bound carp Ig at the cell membrane. P, phagosome; (From Koumans-van Diepen et al., 1994, with permission.)

fish IFN-γ (Novoa et al., 1996). The presence of the macrophage activating factor (MAF) was mentioned above.

Transplantation
Similarly to agnathans and elasmobranch, all teleost species recognize allografts and respond with accelerated second-set memory (Hildemann, 1970), for review see (Manning and Turner, 1976; Manning, 1994). A comparative study used four different species and showed that the acute rejection occurred in all tested species with similar median survival time of 7 days (Ingram, 1980). Generally speaking, graft rejection in osteichtyes is

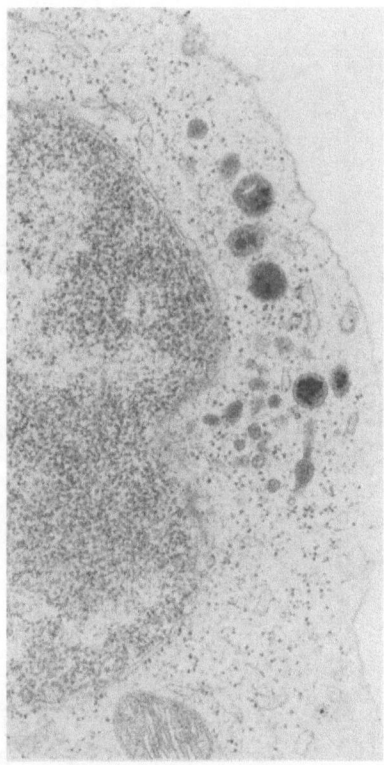

Figure 21. Electron micrograph of a pronephros monocyte-like cell labeled with a monoclonal antibody against Ig (WCI 12). Large gold particles (30 nm) represent internalized Ig and small gold particles (10 nm) represent newly bound carp Ig at the cell membrane. (From Koumans-van Diepen et al., 1994, with permission.)

characterized by a pattern of inflammatory response, lymphocyte infiltration, capillary hemorrhage, and pigment cell destruction (Greenwood et al., 1966).

The eye has been known for a long time to be an organ with so-called immune privilege. Most, if not all the data describing this phenomenon were obtained on mammals. McKinney's group implanted neural retinal or scale allografts and autografts into the vitreous cavity or the anterior chamber of goldfish eye. Neural retinal allografts implanted intraocularly were all rejected by day 8, whereas all autografts survived (Jiang et al., 1994). The authors conclude that the immune privilege in the eye does not exist in goldfish and speculate that it might be an evolutionary adaptation having potential advantages for higher vertebrates.

The experiments evaluating mixed leukocyte reaction (MLR), which is considered to be an *in vitro* analog of the allograft response, are highly controversial. In salmon trout, carp, and bluegill, a positive reaction has been found (Cuchens and Clem, 1977; Etlinger et al., 1977; Caspi and Avtailon, 1984), but in catfish, snappers, and holostean gar, negative results have been obtained (McKinney et al., 1976, 1981) even though these species reject allografts.

Cytotoxicity

Cell-mediated killing in fish belongs to the extremely important features of cellular immunity. Cells of spleen, anterior kidney, thymus and blood were found to induce unrestricted lysis of murine, fish, and human targets (Graves et al., 1984; Hayden and Laux, 1985; Moody et al., 1985), these cells were recently named nonspecific cytotoxic cells (NCC). Besides cytolytic activity oriented against vertebrate targets, catfish NCC lyse also the protozoan fish pathogen, suggesting a broader role in defense system (Jaso-Friedmann et al., 1988). Because of the fact that no allogeneic recognition is involved in NCC action, these cells are considered to be evolutionary progenitors of mammalian natural killer cells. Experiments using trout NCC showed that the target YAC-1 cells were killed by both apoptic and necrotic mechanisms, suggesting functional similarity to vertebrate cytotoxic cells. Light and electronmicroscopic observations (Figs 22 and 23) showed strong target-effector cells binding, the effector cells being small agranular mononuclear lymphocytes (Greenlee et al., 1991). These

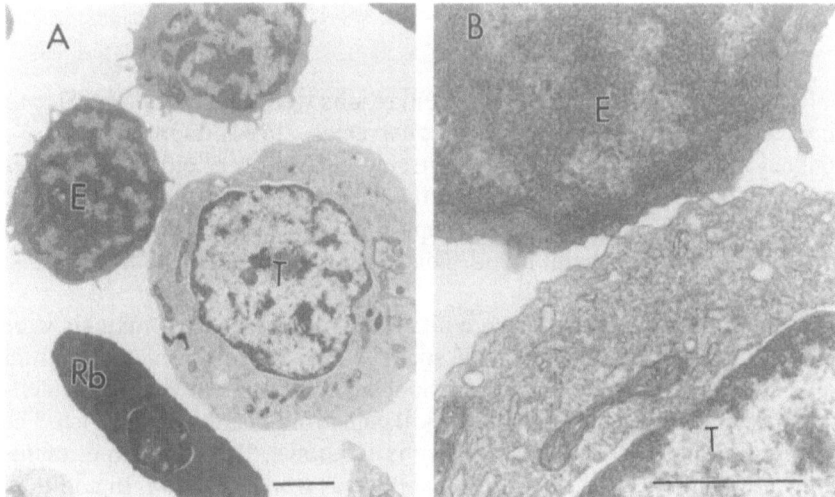

Figure 22. Transmission electron micrographs of trout effector cells (E) conjugating with YAC-1 target cells (T). Effector cells were isolated from the anterior kidney and incubated with target cells at a ratio of 50:1 for 1 h at 15°C and processed for electron microscopy, percent-specific lysis as measured by 4-h DNA fragmentation and chromium release assays were 66.3% and 36.4%, respectively. Rb = red blood cell, (A) 6000 × magnification, shows effector cell with pseudopods and clefted nucleus. (B) 25000 × magnification, illustrates juncture of effector and target cell membranes. Bars represent 1μm. (From Greenlee et al., 1991, with permission.)

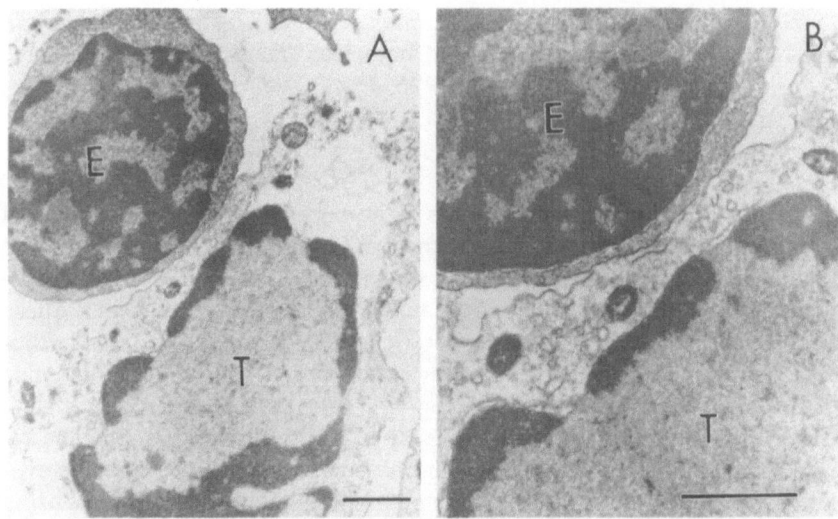

Figure 23. Transmission electron micrographs of trout effector cells (E) conjugating with YAC-1 target cells (T). Effector cells were isolated from the anterior kidney and incubated with target cells at a ratio of 50:1 for 1 h at 15°C and processed for electron microscopy, percent-specific lysis as measured by 4-h chromium release assays was 33.1%. (A) 10000 × magnification, shows condensed and marginated chromatin present in the nucleus of the target cell (consistent with cell death by DNA fragmentation) and swollen mitochondria (consistent with cell death by necrosis). (B) 25000 × magnification, shows interdigitation of effector and target cell membranes. Bars represent 1μm. (From Greenlee et al., 1991, with permission.)

cells are non-adherent. Rather different results were achieved using *Sparus aurata* and *Cyprinus carpio* (Meseguer et al., 1994). Both fish have a leukocyte cell population with ultrastructural features of either monocytes or lymphocytes showing NCC ability. Using isogenic lines of rainbow trout, Ristow et al. found two groups with significantly lower levels of NCC activity in peripheral blood. This low activity appears to be inherited as a recessive trait (1995).

When cells of the damselfish with a malignant neurofibromatosis were studied, a presence of cells with specific anti-neurofibromatosis tumor cells cytotoxicity was found (McKinney and Schmale, 1994). These cells are absent in healthy fish and do not kill cells from healthy damselfish. The cell type responsible for this activity is unknown. Blocking experiments using cold targets were able to block only part of the reaction. In addition, anti-NCC antibody stained only part of the effector cell population. The authors of this study hypothesized that at least two different cell types are involved.

Neutrophils are another cell type involved in spontaneous cytotoxic activity. The sensitive cell lines include K562, HL60, Daudi, HELA and Molt4F; cell lines YAC-1 and P815 were resistant (Kurata et al., 1995).

Inhibitory effects of catalase suggested the involvement of H_2O_2 in cell killing.

In addition to NCC, catfish peripheral blood contains lymphocytes with killer activity against allogeneic cells. These effector cells seem to be distinct from macrophages. The killing was inhibited by monoclonal antibody 1H5 which was suggested to react with fish LFA-1 molecule (Yoshida et al., 1995).

In addition to these two types of cytotoxic cells, Hogan et al. found a third class of effector cells responsible for killing virus-infected autologous and allogeneic cells (Hogan et al., 1996). Inhibitor studies showed that early virus gene products were sufficient to render infected cells susceptible. Inhibition studies and cell depletion studies suggest that two populations of peripheral blood-derived, cytotoxic cells exist in channel catfish—one that lyses virus infected cells, and another that lyses allogeneic cells. In subsequent study, greatly increased cytotoxic responses have been generated by stimulation of lymphocytes with irradiated cells of allogeneic cloned B cell lines in mixed lymphocyte cultures (Stuge et al., 1997). This type of cytotoxicity did not exhibit allospecificity, even though autologous cells were resistant. Despite some unanswered questions such as the origin of these killer cells and the apparent lack of allospecificity, these cells are highly enriched cytotoxic cell population which should facilitate further research efforts.

Similarly to phagocytosis, cell-mediated killing is an especially important defense mechanism for ectothermic vertebrates. The activity of NCC against target cells was enhanced by low (12°C) *in vivo* temperatures; the high (28°C) temperature had no effects (Morvan et al., 1996). The maximum effects occurred on day 28 after exposure and an adaptation in cytotoxic cell activity occurred 56 days after exposure. The inverse effects of temperature on cellular and humoral immunity are often explained as the differences between specific (decreased antibody production) and nonspecific (phagocytosis and cytotoxicity) reactions. On the other hand, the proliferative response of leukocytes to the mitogenic effects of PHA sharply decreased in low temperatures (Le Morvan-Rocher et al., 1995).

Cellular cooperation

Existence of T lymphocytes together with the presence of secondary antibody response have suggested the role for true cooperation between cells involved in defense reactions. One of the proofs is the existence of recognition between T-dependent and T-independent antigens. Hapten-carrier experiments showed that the antibody response to T-dependent antigens requires involvement of both carrier-specific sIg-negative and hapten-specific sIg-positive cells (Yocum et al., 1975). Clem's group devoted their attention to the role of macrophages as the true accessory cells. Their data demonstrated that sIg$^+$ cells responded only to LPS stimulation regardless the presence of macrophages, whereas sIg$^-$ cells required the presence of macrophages for response to either LPS or Con A (Sizemore et al., 1984). Depletion of macrophages resulted in inhibition of response. To respond after challenge with T-independent antigens, cooperation of macrophages and sIg-positive cells was necessary. When T-dependent antigens were used, hapten-specific sIg$^+$ cells, carrier-specific sIg$^-$ cells and macrophages were involved (Miller et al., 1985). In addition, subsequent direct observations in catfish revealed the antigen-presentation of KLH by antigen-pulsed macrophages both in antibody response and in cell proliferation (Vallejo et al., 1990), which further demonstrates the important role of macrophages in both branches of immune responses. Besides macrophages, both monocytes and B lymphocytes were efficient antigen-presenting cells. Furthermore, when monocytes, B cells and long-term monocyte lines have been used as accessory cells presenting well defined protein antigens such as pigeon heart cytochrome C, hen egg lysozyme or horse myoglobin, all these cell types have been found effective and this function has been found to be allogenetically restricted (Vallejo et al., 1991b). The exogenous proteins were actively internalized by antigen-presenting cells. Subcellular fractionation and use of radioactively labeled antigen revealed lysosomal localization. The intensive intracellular traffic of antigeneic fragments between lysosomes, endosomes and outer membrane was found. The use of membrane fractions isolated from antigen-pulsed cells elicited immune response comparable with the use of intact cells. The pretreatment of macrophages with substances known to inhibit antigen processing and/or presentation in mammals strongly blocked the reaction (Vallejo et al., 1992a).

The lower immune response of fish kept at lower temperatures has been well-established. The question remained whether the antigen-presenting function of monocytes/macrophages is responsible for low antibody response or decreased proliferation of T lymphocytes. Clem's group experiments demonstrated that accessory cells incubated with antigen at low temperature, but physiologically still relevant (i.e., 11° and 17°C), elicited the same type of secondary proliferative response, but required a longer (8 h compared to 5 h) incubation. On the other hand, incubation in physiologically irrelevant temperature (such as 4°C) blocked the potentiation of cell proliferation, most probably due to the inhibition of antigen internalization. Identical data were obtained when antibody response was used instead of cell proliferation (Vallejo et al., 1992a, b). The sensitivity of antigen-presentation processes to a single amino acid substitution in a key position of the immunogenic fragment in fish is fully comparable to that in both murine and human systems (Vallejo et al., 1993).

The readers seeking more detailed description of the immune accessory functions in fish should see Šíma and Větvička (1992). Taken together, it would appear that antigen processing and presentation among the diverse taxa of vertebrates are remarkably well conserved. The accessory cell functions are truly comparable, if not exactly identical.

Conclusions

Because bony fish make up a substantial part of human food sources since the Paleocene when genus *Homo* appeared, up to the present, fish immunology is a focus of aquaculture and the fishery industry. The goal of more precise understanding or fish immune reaction is to improve vaccination and preservation of commercially farmed fish against diseases. From evolutionary point of view bony fish are interesting because they are the first vertebrate group the immune strategy of which approaches the more advanced tetrapods. The organization of their lymphoid tissue represents a further advance on the pathway to adaptive immunity. On the other hand, the fish may have retained some fundamental mechanisms of their immunity from the epoch of their common ancestral origin with other jaw vertebrates to the present, which could be important for deeper understanding of immune functions of mammals and humans. One of these features may be the evolutionary experiment to substitute a function of germinal centers by melano-macrophage aggregates and ellipsoids. This solution was nevertheless an evolutionary success. The fish have created an efficient immune system comprising mutually cooperating cell and humoral immunity, including regulatory cytokines or cytokine-like substances. Fish MHC markers seem to be principially similar to those of mammals. Acute rejection of transplanted foreign tissue is accompanied by all indicators of inflammation process and endowed, like the humoral response, by immunological memory. The interesting feature of fish immunity, despite advanced immune pattern, is the presence of only one immunoglobulin class, the tetrameric IgM, which differentiates fish from tetrapods and even of phylogenetically lower sharks (see Chapter on *Chondrichthyes*, this volume). The last but not least significant character of fish is their well-developed lymphatic system which is independent of blood circulation and enables efficient cellular and humoral communications throughout the fish body (for review see Bertin, 1957; Kampmeier, 1969).

References

Agius, C. (1981) Preliminary studies on the ontogeny of the melano-macrophages of teleost haematopoietic tissues and age-related changes. *Dev. Comp. Immunol.* 5:597–606.

Agius, C. and Agbede, S.A. (1984) An electron microscopical study on the genesis of lipofuscin, melanin, and haemosiderin in haemopoietic tissues of fish. *J. Fish Biol.* 24:471–488.

Ainsworth, A.J. (1994) A β-glucan inhibitable zymosan receptor on channel catfish neutrophils. *Vet. Immunol. Immunopathol.* 41:141–152.

Al-Adhami, M.A. and Kunz, Y.W. (1976) Haemopoietic centres in the developing angelfish *Pterophyllum scalare* (Curier and Valenciennes). *Wilhelm Roux's Arch.* 179:393–401.

Anderson, D.P., Roberson, B.S. and Dixon, O.W. (1979) Cellular immune response in rainbow trout *Salmo gairdneri* Richardson to *Yersenia ruckeri* O-antigen monitored by the passive haemolytic plaque assay test, *J. Fish Dis.* 2:169–178.

Andersson, E. and Matsunaga, T. (1993) Complete cDNA sequence of a rainbow trout IgM gene and evolution of vertebrate IgM constant domains. *Immunogenetics* 38:243–250.

Andrew, W. (1959) *Textbook of Comparative Histology*. University Press, New York.

Arkoosh, M.R. and Kaatari, S.L. (1991) Development of immunological memory in rainbow trout (*Oncorhynchus mykiss*). I. An immunochemical and cellular analysis of the B cell response. *Dev. Comp. Immunol.* 15:279–293.

Avtalion, R.R. (1981) Environmental control of the immune response in fish. *CRC Crit. Rev. Environ. Control* 11:163–188.

Avtalion, R.R. and Milgrom, L. (1976) Regulatory effect of temperature and antigen upon immunity in ectothermic vertebrates. I. Influence of hapten density on the immunological and serological properties of penicilloyl-carrier conjugates. *Immunology* 31:589–594.

Belcourt, D.R., Lazure, C. and Bennett, H.P.J. (1993) Isolation and primary structure of the three major forms of granulin-like peptides from hematopoietic tissues of a teleost fish (*Cyprinus carpio*). *J. Biol. Chem.* 268:9230–9237.

Belcourt, D.R., Okawara, Y., Fryer, J.N. and Bennett, H.P. (1995) Immunocytochemical localization of granulin-1 to mononuclear phagocytic cells of the teleost fish *Cyprinus carpio* and *Carassius auratus*. *J. Leukocyte Biol.* 57:94–100.

Bengten, E., Stromberg, S. and Plistrom, L. (1994) Immunoglobulin V-H regions in Atlantic cod (*Gadus morhua* L): their diversity and relationship to V-H families from other species. *Dev. Comp. Immunol.* 18:109–122.

Bernardi, G.B., D'Onofrio, G., Caccio, S. and Bernardi, G. (1993) Molecular phylogeny of bony fishes, based on the amino acid sequence of the growth hormone. *J. Mol. Evol.* 37:644–649.

Bertin, L. (1957) Appareil circulatoire. *In:* P.P. Grassé (ed.): *Traite de Zoologie, Anatomie Systematique Biologie XIII, Agnathes et Poissons: Anatomie Éthologie Systématique*, pp 1400–1459.

Bly, J.E. and Clem, L.W. (1992) Temperature and teleost immune functions. *Fish Shellfish Immunol.* 2:159–171.

Bly, J.E., Buttke, T.M. and Clem, L.W. (1990) Differential effects of temperature and exogenous fatty acids on mitogen-induced proliferation in channel catfish T and B lymphocytes. *Comp. Biochem. Physiol. [A]* 95:417–424.

Bodammer, J.E. (1986) Ultrastructural observations of peritoneal exudate cells from triped bass, *Morone saxatilis*. *Vet. Immunol. Immunopathol.* 12:127–140.

Botham, J.W. and Manning, M.J. (1981) The histogenesis of the lymphoid organs in the carp *Cyprinus carpio* L. and the ontogenetic development of allograft reactivity. *J. Fish Biol.* 19:403–414.

Botham, J.W., Grace, M.F. and Manning, M.J. (1980) Ontogeny of first set and second set alloimmune reactivity in fishes. *In:* M.J. Manning (ed.): *Phylogeny of Immunological Memory*, Elsevier/North Holland Biomedical Press, Amsterdam, pp 83–92.

Bourmaud, C.A.F., Romestand, B. and Bouix, G. (1995) Isolation and partial characterization of IgM-like seabass (*Dicentrarchus labrax* L. 1758) immunoglobulins. *Aquaculture* 132:53–58.

Brattgjerd, S. and Evensen, O. (1996) A sequential light microscopic and ultrastructural study on the uptake and handling of *Vibrio almonicida* in phagocytes of the head kidney in experimentally infected Atlantic salmon (*Salmo salar* L.). *Vet. Pathol.* 33:55–65.

Brown, C.L. and George, C.J. (1985) Age dependent accumulation of macrophage aggregates in the yellow perch *Perca fluviatilis* (Mitchell). *J. Fish Dis.* 8:135–138.

Burnett, K.G. and Schwartz, L.K. (1994) Leukocyte proliferation mediated by protein kinase C in the marine teleost fish, *Sciaenops ocellatus*. *Dev. Comp. Immunol.* 18:33–43.

Carroll, R.L. (1988) *Vertebrate Paleontology and Evolution*. W.H. Freeman and Co., New York.

Caspi, R.R. and Avtalion, R.R. (1984) The mixed leukocyte reaction (MLR) in carp: bidirectional and unidirectional MLR response. *Dev. Comp. Immunol.* 8:631–637.

Chandler, A.C. (1911) On a lymphoid structure lying over the myelencephalon of *Lepisosteus*. *Univ. California Publ. Zool.* 9:85–104.

Chiller, J.M., Hodgins, H.O. and Welser, R.S. (1969) Antibody response in rainbow trout (*Salmo gairdneri*). II. Studies in the kinetics of development of antibody-producing cells and on complement and natural hemolysin. *J. Immunol.* 102:1202–1207.

Chilmonczyk, S. (1983) The thymus of the rainbow trout *Salmo gairdneri*. Light and electron microscopic study. *Dev. Comp. Immunol.* 7:59–68.

Clem, L.W., Miller, N.W. and Bly, J.E. (1991) Evolution of lymphocyte subpopulations: their interactions and temperature sensitivities. *In:* G.W. Warr and N. Cohen (eds): *Phylogenesis of Immune Functions,* CRC Press, Boca Raton, pp 191–213.

Collazos, M.E., Barriga, C. and Ortega, E. (1995) Seasonal changes in phagocytic capacity and superoxide anion production of blood phagocytes from tench (*Tinca tinca* L). *J. Comp. Physiol.* 165:71–76.

Cooper, E.L., Zapata, A., Garcia Barrutia, M. and Ramirez, J.A. (1983) Aging changes in lymphopoietic and myelopoietic organs of the annual cyprinodont fish, *Nothoranchius guentheri. Exp. Gerontol.* 18:29–38.

Cuchens, M.A. and Clem, L.W. (1977) Phylogeny of lymphocyte heterogeneity. II. Differential effects of temperature on fish T-like and B-like cells. *Cell. Immunol.* 34:219–230.

Cuchens, M.A., McLean, E. and Clem, L.W. (1976) Lymphocyte heterogeneity in fish and reptiles. *In:* R.K. Wright and E.L. Cooper (eds): *Phylogeny of Thymus and Bone Marrow-Bursa Cells,* Elsevier/North Holland Biomedical Press, Amsterdam, pp 205–213.

Dalmo, R.A., Ingebrigtsen, K., Sveinbjornsson, B. and Seljelid, R. (1966) Accumulation of immunomodulatory laminaran ($\beta(1,3)$-D-glucan) in the heart, spleen and kidney of Atlantic cod, *Gadus morhua* L. *J. Fish Dis.* 19:129–136.

Davina, J.H.M., Rijkers, G.T., Rombout, J.H.W.M., Timmermans, L.P.M. and van Muiswinkel, W.B. (1980) Lymphoid and non-lymphoid cells in the intestine of cyprinid fish. *In:* J.D. Horton (ed.): *Development and Differentiation of Vertebrate Lymphocytes,* Elsevier/North Holland Biomedical Press, Amsterdam, pp 129–240.

Desvaux, F.X. and Charlemagne, J. (1981) The goldfish immune response. I. Characterization of the humoral response to particulate antigens. *Immunology* 43:621–627.

Dexiang, C. and Ainsworth, A.J. (1991) Assessment of metabolic activation of channel catfish peripheral blood neutrophils. *Dev. Comp. Immunol.* 15:201–208.

Dixon, B., Vanerp, S.H.M., Rodrigues, P.N.S., Egberts, E. and Stet, R.J.M. (1995) Fish major histocompatibility complex genes: an expansion. *Dev. Comp. Immunol.* 19:109–133.

Dorin, D., Sire, M.F. and Vernier, J.M. (1993) Endocytosis and intracellular degradation of heterologous protein by eosinophilic granulocytes isolated from rainbow trout (*Oncorhynchus mykiss*) posterior intestine. *Biol. Cell* 79:219–224.

Ellis, A.E. (1974) Aspects of the lymphid and reticuloendothelial systems in the plaice, *Pleuronectes platessa* L. Ph.D. Thesis, University of Aberdeen.

Ellis, A.E. (1977) Ontogeny of the immune response in *Salmo salar.* Histogenesis of lymphoid organs and appearance of membrane immunoglobulin and mixed leucocyte reactivity. *In:* J.B. Solomon and J.D. Horton (eds): *Developmental Immunobiology,* Elsevier/North Holland Biomedical Press, Amsterdam, pp 225–231.

Ellis, A.E. (1980) Antigen trapping in the spleen and kidney of the plaice, *Pleuronectes platessa. J. Fish Dis.* 3:413–426.

Ellis, A.E. (1981) Non-specific defense mechanisms in fish and their role in disease proceses. *Develop. Biol. Stand.* 49:337–352.

Ellis, A.E. (1982) Differences between the immune mechanisms of fish and higher vertebrates. *In:* R.J. Roberts (ed.): *Microbial Diseases in Fish,* Academic Press, London, pp 1–29.

Ellis, A.E. and de Sousa, M.A.B. (1974) Phylogeny of the lymphoid system 1. A study of the fate of circulating lymphocytes in plaice. *Eur. J. Immunol.* 4:338–343.

Ellis, A.E., Munroe, A.L.S. and Roberts, R.J. (1976) Defense mechanisms in fish. I. A study of the phagocytic system and the fate of intraperitoneally injected particulate material in the plaice *Pleuronectes platessa. J. Fish. Biol.* 8:68–78.

Ellsaesser, C.F. and Clem, L.W. (1994) Functionally distinct high and low molecular weight species of channel catfish and mouse IL-1. *Cytokine* 6:10–20.

Ellsaesser, C.F., Miller, N.W., Cuchens, M.A., Lobb, C.J. and Clem, L.W. (1985) Analysis of catfish peripheral leucocytes by brightfield microscopy and flow cytometry. *Trans. Am. Fish Soc.* 114:279–285.

Ellsaesser, C.F., Bly, J.E. and Clem, L.W. (1988) Phylogeny of lymphocyte heterogeneity: the thymus of the channel catfish. *Dev. Comp. Immunol.* 12:787–799.

Engstad, R.E. and Robertsen, B. (1994) Specificity of a β-glucan receptor on macrophages from Atlantic salmon (*Salmo salar* L.). *Dev. Comp. Immunol.* 18:397–408.

Estevez, J., Leiro, J., Sanmartin, M.L. and Ubeira, F.M. (1993) Isolation and partial characterization of turbot (*Scophthalmus maximus*) immunoglobulins. *Comp. Biochem. Physiol. [A]* 105:275–281.

Etlinger, H.M., Hodgins, H.O. and Chiller, J.M. (1976) Evolution of the lymphoid system. I. Evidence for lymphocyte heterogeneity in rainbow trout revealed by the organ distribution of mitogenic responses. *J. Immunol.* 166:1547–1553.

Etlinger, H.M., Hodgins, H.O. and Chiller, J.M. (1977) Evolution of the lymphoid system. II. Evidence for immunoglobulin determinants on all rainbow trout leukocytes and demonstration of mixed leukocyte reaction. *Eur. J. Immunol.* 7:881–887.

Fänge, R. and Pulsdorf, A. (1985) The thymus of the angler fish *(Lophius piscatorius* [Pisces: Teleostei]): A light and electron microscopic study, *In:* M.J. Manning and M.F. Tatner (eds): *Fish Immunology,* Academic Press, London, pp 293–311.

Ferren, F.A. (1967) Role of the spleen in the immune response of teleosts and elasmobranchs. *J. Florida Med. Assoc.* 54:434–437.

Fichtelius, K.E., Finstad, J. and Goodm R.A. (1968) Bursa equivalents of bursaless vertebrates. *Lab. Invest.* 19:339–351.

Gage, L.A. (1976) *Wildlife Diseases.* Plenum Press, New York.

Georgopolou, U. and Vernier, L.M. (1986) Local immunological response in the posterior intestinal segment of the rainbow trout after oral administration of macromolecules. *Dev. Comp. Immunol.* 10:529–537.

Ghaffari, S.H. and Lobb, C.J. (1989) Cloning and sequence analysis of channel catfish heavy chain cDNA indicate phylogenetic diversity within the IgM immunoglobulin family. *J. Immunol.* 142:1356–1365.

Ghaffari, S.H. and Lobb, C.J. (1993) Structure and genomic organization of immunoglobulin light chain in the channel catfish. An unusual genomic organizational pattern of segmental genes. *J. Immunol.* 151:6900–6912.

Ghaffari, S.H. and Lobb, C.J. (1997) Structure and genomic organization of a second class of immunoglobulin light chain genes in the channel catfish. *J. Immunol.* 159: 250–258.

Godwin, U.B., Antao, A., Wilson, M.R., Chinchar, V.G., Miller, N.W., Clem, L.W. and McConnell, T.J. (1997) MHC class II *B* genes in the channel catfish *(Ictalurus punctatus). Dev. Comp. Immunol.* 21:13–23.

Gorgollon, P. (1983) Fine structure of the thymus in the adult cling fish *Sicyastes sanguineus* (Pisces, Gobiesocidae). *J. Morphol.* 177:25–40.

Good, R.A., Finstad, J., Pollara, B. and Gabrielsen, A.E. (1966) Morphological studies on the evolution of the lymphoid tissues among the lower vertebrates. *In:* R.T. Smith, P.A. Miescher and R.A. Good (eds): *Phylogeny of Immunity,* University of Florida Press, Gainesville, pp 149–168.

Grace, M.F. and Manning, M.J. (1980) Histogenesis of the lymphoid organs n rainbow trout *Salmo gairdneri. Dev. Comp. Immunol.* 4:255–264.

Graf, R. and Schluns, J. (1979) Ultrastructural and histochemical investigation of the terminal capillaries of the spleen of the carp *Cyprinus carpio* L. *Cell Tissue Res.* 196:289–306.

Graham, S. and Secombes, C.J. (1990) Do fish lymphocytes secrete interferon-gamma? *J. Fish Biol.* 36:563–573.

Graser, R., Ohuigin, C., Vincek, V., Meyer, A. and Klein, J. (1996) Trans-species polymorphism of class II Mhc loci in danio fishes. *Immunogenetics* 44:36–48.

Graves, S.S., Evans, D.L., Cobb, D. and Dawe, D.L. (1984) Nonspecific cytotoxic cells of fish *(Ictalurus punctatus).* I. Optimum requirements for target cell lysis. *Dev. Comp. Immunol.* 8: 293–302.

Greenlee, A.R., Brown, R.A. and Ristow, S.S. (1991) Nonspecific cytotoxic cells of rainbow trout *(Oncorhynchus mykiss)* kill YAC-1 targets by both necrotic and apoptic mechanisms. *Dev. Comp. Immunol.* 15:153–164.

Greenwood, P.H., Rosen, D.E., Wertman, S.H. and Meyers, G.S. (1966) Phyletic studies of teleostean fishes with a provisional classification of living forms. *Bull. Am. Mus. Natl. Hist.* 131:399–431.

Grimholt, U., Hordvik, I., Fosse, V.M., Olsaker, I., Endresen, C. and Lie, O. (1993) Molecular cloning of major histocompatibility complex class I cDNAs from Atlantic salmon *(Salmo salar). Immunogenetics* 37:469–473.

Grondel, J.L. and Harmsen, E.G.M. (1984) Phylogeny of interleukins: growth factors produced by leucocytes of the cyprinid fish, *Cyprinus carpio* L. *Immunology* 52:477–482.

Hafter, E. (1952) Histological age changes in the thymus of the teleost *Astyanax. J. Morphol.* 90:551–581.

Hamby, B.A., Huggins, E.M., Lachman, L.B., Dinarello, C.A. and Sigel, M.M. (1986) Fish lymphocytes respond to human IL-1. *Lymphokine Res.* 5:157–162.

Hardie, L.J., Fletcher, T.C. and Secombes, C.J. (1994) Effect of temperature on macrophage activation and the production of macrophage activating factor by rainbow trout (*Oncorhynchus mykiss*) leukocytes. *Dev. Comp. Immunol.* 18:57–66.

Hashimoto, K., Nakanishi, T. and Kurosawa, Y. (1990) Isolation of carp genes encoding major histocompatibility complex antigens. *Proc. Natl. Acad. Sci. U.S.A.* 87:6863–6867.

Hayden, B.J. and Laux, D.C. (1985) Cell-mediated lysis of murine target cells by nonimmune salmonid lymphoid preparations. *Dev. Comp. Immunol.* 9:627–639.

Herraez, M.P. and Zapata, A. (1986) Structure and function of the melano-macrophage centres of the goldfish *Carassius auratus. Vet. Immunol. Immunopathol.* 12:117–126.

Hildemann, W.H. (1970) Transplantation immunity in fishes: Agnatha, Chondrichthyes and Osteichthyes. *Transplant. Proc.* 11:253–259.

Hildemann, W.H. and Haas, R. (1960) Comparative studies of homotransplantation in fishes. *J. Cell. Comp. Physiol.* 55:227

Hogan, R.J., Stuge, T.B., Clem, L.W., Miller, N.W. and Chinchar, V.G. (1996) Anti-viral cytotoxic cells in the channel catfish (*Ictalurus punctatus*). *Dev. Comp. Immunol.* 20:115–127.

Hordvik, I., Voie, A.M., Glette, J., Male, R. and Endresen, C. (1992) Cloning and sequence analysis of two isotypic IgM heavy chain genes from Atlantic salmon, *Salmo salar* L. *Eur. J. Immunol.* 22:2957–2962.

Hrubec, T.C., Robertson, J.L., Smith, S.A. and Tinker, M.K. (1996) The effect of temperature and water quality on antibody response to *Aeromonas salmonicida* in sunshine bass (*Morone chrysops* X *Morone saxatilis*). *Vet. Immunol. Immunopathol.* 50:157–166.

Imagawa, T., Hashimoto, Y., Kon, Y. and Sugimura, M. (1991) Immunoglobulin containing cells in the head kidney of carp (*Cyprinus carpio* L.) after bovine serum albumin injection. *Fish Shellfish Immunol.* 1:173–185.

Ingram, G.A. (1980) Substances involved in natural resistance of fish to infection. A review. *J. Fish Biol.* 16:23–60.

Jang, S.I., Hardie, L.J. and Secombes, C.J. (1994) Effects of transforming growth factor β_1 on rainbow trout *Oncorhynchus mykiss* macrophage respiratory burst activity. *Dev. Comp. Immunol.* 18:315–323.

Jaso-Friedmann, L., Evans, D.L., Grant, C.C., St.John, A., Harris, D.T. and Koren, H.S. (1988) Characterization by monoclonal antibodies of a target cell antigen complex recognitized by nonspecific cytotoxic cells. *J. Immunol.* 141:2861–2868.

Jensen, L.E., Petersen, T.E., Thiel, S. and Jensenius, J.C. (1995) Isolation of a pentraxin-like protein from rainbow trout serum. *Dev. Comp. Immunol.* 19:305–314.

Jensen, L.E., Hiney, M.P., Shields, D.C., Uhlar, C.M., Lindsay, A.J. and Whitehead, A.S. (1997a) Acute phase protein in salmonids. Evolutionary analyses and acute phase response. *J. Immunol.* 158:384–392.

Jensen, L.E., Thiel, S., Petersen, T.E. and Jensenius, J.C. (1997b) A rainbow trout lectin with multimeric structure. *Comp. Biochem. Physiol. [B]* 116:385–390.

Jiang, L.Q., Streilein, J.W. and McKinney, C. (1994) Immune privilege in the eye: an evolutionary adaptation. *Dev. Comp. Immunol.* 18:421–431.

Jorgensen, J.B., Lunde, H. and Robertsen, B. (1993) Peritoneal and head kidney cell response to intraperitoneally injected yeast glucan in Atlantic salmon, *Salmo salar* L. *J. Fish Dis.* 16:313–325.

Jorgensen, J.B. and Robertsen, B. (1995) Yeast β-glucan stimulates respiratory burst activity of atlantic salmon (*Salmo salar* L.) macrophages. *Dev. Comp. Immunol.* 19:43–57.

Juul-Madsen, H.R., Glamann, J., Madsen, H.O. and Simonsen, M. (1992) MHC class II beta-chain expression in the rainbow trout. *Scad. J. Immunol.* 35:687–694.

Kampmeier, O.F. (1969) Lymphatic system of the bony fish. In: *Evolution and Comparative Morphology of the Lymphatic System,* Charles T. Thomas, Springfield, pp 232–265.

Killie, J.E.A. and Jorgensen, T.O. (1994) Immunoregulation in fish. I. Intramolecular-induced suppression of antibody responses to haptenated protein antigens studied in Atlantic salmon (*Salmo salar* L.). *Dev. Comp. Immunol.* 18:123–136.

Killie, J.E.A. and Jorgensen, T.O. (1995) Immunoregulation in fish. II. Intermolecular-induced suppression of antibody responses studied by haptenated antigens in Atlantic salmon (*Salmo salar* L.). *Dev. Comp. Immunol.* 19:389–404.

Kimura, N. and Kudo, S. (1975) Fine structure of the stratum granulosum in the pyloric caeca of the rainbow trout. *Jap. J. Ichthyol.* 22:16–22.

Kodama, H., Hirota, Y., Mukamoto, M., Baba, T. and Azuma, I. (1993) Activation of rainbow trout (*Oncorhynchus mykiss*) phagocytes by muramyl dipeptide. *Dev. Comp. Immunol.* 17: 129–140.

Koumans-van Diepen, J.C.E., van de Lisdonk, M.H.M., Taverne-Thiele, A.J., Verburg-van Kemenade, B.M.L. and Rombout, J.H.W.M. (1994) Characterization of immunoglobulin-binding leucocytes in carp (*Cyprinus carpio* L.). *Dev. Comp. Immunol.* 18:45–56.

Koumans-van Diepen, J.C.E., Egberts, E., Peixoto, B.R., Taverne, N. and Rombout, J.H.W.M. (1995) B cell and immunoglobulin heterogeneity in carp (*Cyprinus carpio L*): an immuno(cyto)chemical study. *Dev. Comp. Immunol.* 19:97–108.

Kunich, J.C., Contreras, B., Recktenwald, D.J. and Nieto, M.C. (1994) A demonstration of human beta-2-microglobulin specific association with the surface of teleostean cells. *Dev. Comp. Immunol.* 18:483–494.

Kurata, O., Okamoto, N. and Ikeda, Y. (1995) Neutrophilic granulocytes in carp, *Cyprinus carpio*, possess a spontaneous cytotoxic activity. *Dev. Comp. Immunol.* 19:315–325.

Lauder, G.V. and Liehm, K.F. (1983) The evolution and interrelationships of the actinopterygian fishes. *Bull. Museum Comp. Zool.* 150:95–197.

Le Morvan-Rocher, C., Troutaud, D. and Deschaux, P. (1995) Effects of temperature on carp leukocyte mitogen-induced proliferation and nonspecific cytotoxic activity. *Dev. Comp. Immunol.* 19:87–95.

Lin, G.L., Ellsaesser, C.F., Clem, L.W. and Miller, N.W. (1992) Phorbol ester/calcium iono-phore activate fish leukocyte and induce long-term cultures. *Dev. Comp. Immunol.* 16: 153–163.

MacAllister, D.E. (1987) *A Working List of the Fishes of the World*. Ichthyology Section, National Museum of Natural Sciences, National Museums of Canada, Ottawa.

Manning, M.J. (1981) The evolution of vertebrate lymphoid organs. *In:* J.B. Solomon (ed.): *Aspects of Developmental and Comparative Immunology I*, Pergamon Press, Oxford, pp 67–72.

Manning, M.J. (1994) Fishes. *In:* R.J. Turner (ed.): *Immunology: A comparative Approach*, John Wiley & Sons, Chichester, pp 69–100.

Manning, M.J. and Turner, R.J. (1976) *Comparative Immunobiology*. Blackie, Glasgow.

Matsunaga, T., Chen, T. and Tormanen, V. (1990) Characterization of a complete immunoglo-bulin heavy-chain variable region germ-line gene of rainbow trout. *Proc. Natl. Acad. Sci. U.S.A.* 87:7767–7771.

May, R.T. (1990) How many species? *Phil. Trans. R. Soc. London B* 330:293–304.

McCumber, L.J., Sigel, M.M., Trauger, R.J. and Cuchens, M.A. (1982) RES structure and func-tion of the fishes. *In:* N. Cohen and M.M. Sigel (eds): *The Reticuloendothelial System, Vol. 3.*, Plenum Press, New York, pp 393–422.

McKinney, E.C., McLeod, T.F. and Sigel, M.M. (1981) Allograft rejection in a holostean fish, *Lepisosteus platyrinchus*. *Dev. Comp. Immunol.* 5:65–74.

McKinney, E.C., Ortiz, G., Lee, J.C., Sigel, M.M., Lopez, D.M., Epstein, R.S. and McLeod, T.F. (1976) Lymphocytes of fish: multipotential or specialized? *In:* R.K. Wright and E.L. Cooper (eds): *Phylogeny of Thymus and Bone Marrow Bursa Cells*, Elsevier/North-Holland Biome-dical Press, Amsterdam, pp 73–82.

McKinney, E.C. and Schmale, M.C. (1994) Damselfish with neurofibromatosis exhibit cyto-toxicity toward tumor targets. *Dev. Comp. Immunol.* 18:305–313.

Meseguer, J., Esteban, M.A., Lopezruiz, A. and Bielek, E. (1994) Ultrastructure of nonspecific cytotoxic cells in teleost. 1. Effector-target cell binding in a marine and freshwater species (seabream: *Sparus aurata* L., and carp: *Cyprinus carpio* L.). *Anat. Rec.* 239:468–474.

Miller, N.W. and Clem, L.W. (1984) Microsystem for *in vitro* primary and secondary immun-ization of channel catfish (*Ictalurus punctatus*) leucocytes with hapten-carrier conjugates. *J. Immunol. Methods*, 72:367–379.

Miller, N.W., Sizemore, R.C. and Clem, L.W. (1985) Phylogeny of lymphocyte heterogeneity: the cellular requirements for *in vitro* antibody response of channel catfish lymphocytes. *J. Immunol.* 134:2884–2888.

Miller, N.W., Rycyzyn, M.A., Wilson, M.R., Warr, G.W., Naftilan, A.J. and Clem, L.W. (1994) Development and characterization of channel catfish long term B cell lines. *J. Immunol.* 152: 2180–2189.

Mochida, K., Lou, Y.H., Hara, A. and Yamauchi, K. (1994) Physical biochemical properties of IgM from a teleost fish. *Immunology* 83:675–680.

Moody, C.E., Serreze, D.V. and Reno, P.W. (1985) Nonspecific cytotoxic activity of teleost leukocytes. *Dev. Comp. Immunol.* 9:51–64.

Morvan, C.L., Deschaux, P. and Troutaud, D. (1996) Effects and mechanisms of environmental temperature on carp *(Cyprinus carpio)* anti-DNP antibody response and non-specific cytotoxic cell activity: a kinetic study. *Dev. Comp. Immunol.* 20:331–340.

Nagae, M., Fuda, H., Hara, A., Kawamura, H. and Yamauchi, K. (1993) Changes in serum immunoglobulin M (IgM) concentrations during early development of chum salmon *(Oncorhynchus keta)* as determined by sensitive ELISA technique. *Comp. Biochem. Physiol. [A]* 106:69–74.

Nakamura, H. and Shimozawa, A. (1994) Phagocytic cells in the fish heart. *Arch. Histol. Cytol.* 57:415–425.

Nakamura, H., Furuta, E. and Shimozawa, A. (1992) *In vivo* response to administred carbon particles in the teleost, *Oryzias latipes. Dokkyo J. Med. Sci.* 19:11–18.

Nakamura, H., Shimozawa, A. and Kikuchi, S.I. (1993) Melano-macrophage centre-like structures in the heart of the medaka, *Oryzias latipes. Ann. Anat.* 175:59–63.

Nakao, M., Uemura, T. and Yano, T. (1996) Terminal components of carp complement constituing a membrane attack complex. *Mol. Immunol.* 33:933–937.

Nelson, J.S. (1984) *Fishes of the World.* John Wiley and Sons, New York.

Normark, B.B., McCune, A.R. and Harrison, R.G. (1991) Phylogenetic relationships of neopterygian fishes inferred from mitochondrial DNA sequences. *Mol. Biol. Evol.* 8: 819–834.

Neumann, N.F. and Belosevic, M. (1996) Deactivation of primed respiratory burst response of goldfish macrophages by leukocyte-derived macrophage activating factor(s). *Dev. Comp. Immunol.* 30:427–439.

Neumann, N.F., Fagan, D. and Belosevic, M. (1995) Macrophage activating factor(s) secreted by mitogen stimulated goldfish kidney leukocytes synergize with bacterial lipopolysaccharide to induce nitric oxide production in teleost macrophages. *Dev. Comp. Immunol.* 19: 473–482.

Nonaka, M., Natsuume-Sakai, S. and Takahashi, M. (1991) The complement system in rainbow rout *(Salmo gairdneri)*. II. Purification and characterization of the fifth component (C5). *J. Immunol.* 126:1495–1498.

Novoa, B., Figueras, A., Ashton, I. and Secombes, C.J. (1996) *In vitro* studies on the regulation of rainbow trout *(Oncorhynchus mykiss)* macrophage respiratory burst activity. *Dev. Comp. Immunol.* 20:207–216.

Ono, H., Klein, D., Vincek, V., Figueroa, F., O'hUigin, C., Tichy, H. and Klein, J. (1992) Major histocompatibility complex class II genes in zebrafish. *Proc. Natl. Acad. Sci. U.S.A.* 89: 11886–11890.

Partula, S., Fellah, J.S., de Guerra, A. and Charlemagne, J. (1994) Characterization of cDNA of T cell receptor beta chain in rainbow trout. *Compt. Rendus Acad. Sci.* 317:765–770.

Partula, S., Deguerra, A., Fellah, J.S. and Charlemagne, J. (1995) Structure and diversity of the T cell antigen receptor beta-chain in a teleost fish. *J. Immunol.* 155:699–706.

Passer, B.J., Chen, C.H., Miller, N.W. and Cooper, M.D. (1996) Identification of a T lineage antigen in the catfish. *Dev. Comp. Immunol.* 20:441–450.

Patterson, C. and Rosen, D. (1977) Review of the ichthyodectiform and other Mesozoic teleost fishes and the theory and practice of classifying fossils. *Bull. Am. Museum Nat. Hist.* 159: 81–172.

Pedrera, I.M., Collazos, M.E., Ortega, E. and Barriga, C. (1992) *In vitro* study of the phagocytic processes in splenic granulocytes of the tench *(Tinca tinca, L.). Dev. Comp. Immunol.* 16: 431–439.

Peleteiro, M.C. and Richards, R.H. (1985) Identification of lymphocytes in the epidermis of the rainbow trout, *Salmo gairdneri* Richardson. *J. Fish Dis.* 8:161–172.

Pilstrom, L. and Petersson, A. (1991) Isolation and partial characterization of immunoglobulin from cod *(Gadus morhua* L.). *Dev. Comp. Immunol.* 15:143–152.

Pontius, H. and Ambrosius, H. (1972) Beiträge zur Immunobiologie poikilothermer Wirbeltiere. IX. Untersuchungen zur zellulären Grundlage humoraler Immunoreaktionen der Knochenfische am Beispiel des Flussbarsches *(Perca fluviatilis* L.). *Acta Biol. Med. Germ.* 29:319–339.

Plytycz, B. and Jozkowicz, A. (1994) Differential effects of temperature on macrophages of ectothermic vertebrates. *J. Leukocyte Biol.* 56:729–731.

Rijkers, G.T., Freederix-Wolters, E.M.H. and von Muiswinkel, W.B. (1980a) The effect of antigen dose and route of administration on the development of immunological memory in carp (*Cyprinus carpio*). *In:* M.J. Manning (ed.): *Phylogeny of Immunological Memory,* Elsevier/North Holland Biomedical Press, Amsterdam, pp 99–102.

Rijkers, G.T., Freederix-Wolters, E.M.H. and von Muiswinkel, W.B. (1980b) The immune-system of cyprinid fish. Kinetics and temperature dependence of antibody-producing cells in carp (*Cyprinus carpio*). *Immunology* 41:91–97.

Ristow, S.S., Grabowski, L.D., Wheeler, P.A., Prieur, D.J. and Thorgaard, G.H. (1995) Arlee line of rainbow trout (*Oncorhynchus mykiss*) exhibit a low level of nonspecific cytotoxic cell activity. *Dev. Comp. Immunol.* 19:497–505.

Rodrigues, P.N.S., Hermsen, T.T., Rombout, J.H.W.M., Egberts, E. and Stet, R.J.M. (1995) Detection of MHC class II transcripts in lymphoid tissues of the common carp (*Cyprinus carpio* L.). *Dev. Comp. Immunol.* 19:483–496.

Rombout, J.H.W.M. and van den Berg, A.A. (1989) Immunological importance of the second gut segment of carp. I. Uptake and processing of antigens by epithelial cells and macrophages. *J. Fish Biol.* 35:13–22.

Rombout, J.H.W.M., Blok, L.J., Lamers, C.H.J. and Egberts, E. (1986) Immunization of carp (*Cyprinus carpio*) with *Vibrio anguillarum* bacteria: Indications for a common mucosal immune system. *Dev. Comp. Immunol.* 10:341–352.

Rombout, J.H.W.M., Bot, H.E. and Taverne-Thiele, J.J. (1989) Immunological importance of the second gut segment of carp. II. Characteristics of mucosal leucocytes. *J. Fish Biol.* 35:167–178.

Rombout, J.H.W.M., van de Wall, J.W., Companjen, A., Taverne, N. and Taverne-Thiele, J.J. (1997) Characterization of a T-cell lineage marker in carp (*Cyprinus carpio* L.). *Dev. Comp. Immunol.* 21:35–46.

Rowley, A.F., Hunt, T.C., Page, M. and Mainwaring, G. (1988) Fish. *In:* A.F. Rowley and N.A. Ratcliffe (eds): *Vertebrate Blood Cells,* Cambridge Univ. Press, Cambridge, pp 19–22.

Rosen, D. (1982) Teleostean interrelationships, morphological function, and evolutionary inference. *Am. Zool.* 22:261–273.

Rubem, L.N., Warr, G.W., Decker, J.M. and Marchalonis, J.J. (1977) Phylogenetic origin of immune recognition: Lymphoid heterogeneity and the hapten/carrier effect in goldfish *Carassius auratus. Cell. Immunol.* 31:266–283.

Rycyzyn, M.A., Wilson, M.R., Warr, G.W., Clem, L.W. and Miller, N.W. (1996) Membrane immunoglobulin-associated molecules on channel catfish B lymphocytes. *Dev. Comp. Immunol.* 20:341–251.

Saha, K., Dash, K. and Sahu, A. (1993) Antibody dependent haemolysin, complement, and opsonin in sera of a major carp, *Cirrhina mrigala* and catfish, *Clarias batrachus* and *Heteropneustes fossilis. Comp. Immun. Microbiol. Infect. Dis.* 16:323–330.

Sailendri, K. (1973) Studies on the development of lymphoid organs and immune responses in the teleost *Tilapia mossambica* (Peters). Doctoral Thesis, Madurai University.

Sailendri, K. and Muthukkarruppan, V.R. (1975a) Morphology of lymphoid organs in a cichlid teleost *Tilapia mossambica* (Peters). *J. Morphol.* 147:109–122.

Sailendri, K. and Muthukkarruppan, V.R. (1975b) The immune response of the teleost, *Tilapia mossambica* to soluble and cellular antigens. *J. Exp. Zool.* 191:371–382.

Sakai, D.K. (1984) Opsonization by fish antibody and complement in the immune phagocytosis by peritoneal exudate cells isolated from salmoid fishes. *J. Fish Dis.* 7:29–38.

Sato, K., Figueroa, F., O'hUigin, C., Reznick, D.N. and Klein, J. (1995) Identification of major histocompatibility complex genes in the guppy, *Poecilia reticulata. Immunogenetics* 43:38–49.

Scharrer, E. (1944) The histology of the meningeal myeloid tissue in the ganoids *Amia* and *Lepisosteus. Anat. Rec.* 88:291–310.

Secombes, C.J. and Manning, M.J. (1980) Comparative studies on the immune system of fishes and amphibians. I. Antigen localization in the carp *Cyprinus carpio. J. Fish Dis.* 3:399–412.

Secombes, C.J., Manning, M.J. and Ellis, A.E. (1982a) Localization of the immune complexes and heat-aggregated immunoglobulin in the carp *Cyprinus carpio* L. *Immunology* 47:101–105.

Secombes, C.J., Manning, M.J. and Ellis, A.E. (1982b) The effect of primary and secondary immunization on the lymphoid tissues of carp *Cyprinus carpio* L. *J. Exp. Zool.* 220:277–287.

Secombes, C.J., White, A., Fletcher, T.C. and Houlihan, D.F. (1991) The development of an ELISPOT assay to quantify total and specific antibody-secreting cells in dab *Limanda limanda* (L.). *Fish Shellfish Immunol.* 1:87–97.

Sharp, G.J.E., Pettit, T.R., Rowley, A.F. and Secombes, C.J. (1992) Lipoxin-induced migration of fish leukocytes. *J. Leukocyte Biol.* 51:140–145.

Sigel, M.M., Hamby, B.A. and Huggins, E.M. (1986) Phylogenetic studies on lymphokines. Fish lymphocytes respond to human IL-1 and epithelial cells produce IL-1 like factor. *Vet. Immunol. Immunopathol.* 12:47–58.

Šíma, P. and Větvička, V. (1990) *Evolution of immune reactions.* CRC Press, Boca Raton.

Šíma, P. and Větvička, V. (1992) Evolution of immune accessory functions. *In:* L. Fornůsek and V. Větvička (eds): *Immune System Accessory Cells,* CRC Press, Boca Raton, pp 1–55.

Sizemore, R.C., Miller, N.W., Cuchens, M.A., Lobb, C.J. and Clem, L.W. (1984) Phylogeny of lymphocyte heterogeneity: the cellular requirements for *in vitro* mitogenic responses of channel catfish leukocytes. *J. Immunol.* 133:2920–2924.

Smith, A.S., Potter, M. and Merchant, E.B. (1967) Antibody-forming cells in the pronephros of the teleost, *Lepomis macrochirus. J. Immunol.* 99:876–882.

Smith, A.M., Wivel, N.A. and Potter, M. (1970) Plasmacytosis in the pronephros of the carp *Cyprinus carpio. Anat.Rec.* 167:351–370.

St. Louis-Cormier, E.A., Osterland, C.K. and Anderson, P.D. (1984) Evidence for a cutaneous secretory immune system in rainbow trout (*Salmo gaidneri*). *Dev. Comp. Imunol.* 8:71–80.

Stuge, T.B., Yoshida, S.H., Chinchar, V.G., Miller, N.W. and Clem, L.W. (1997) Cytotoxic activity generated from channel catfish peripheral blood leukocytes in mixed leukocyte cultures. *Cell. Immunol.* 177:154–161.

Sultmann, H. Mayer, W.E., Figueroa, F. Ohuigin, C. and Klein, J. (1993) Zebrafish Mhc class II alpha chain-encoding genes: polymorphism, expression, and function. *Immunogenetics* 38:408–420.

Sultmann, H., Mayer, W.E., Figueroa, F., Ohuigin, C. and Klein, J. (1994) Organization of Mhc class II B genes in the zebrafish (*Brachydanio rerio*). *Genomics* 23:1–14.

Sunyer, J.O., Tort, L. and Lambris, J.D. (1997) Structural C3 diversity in fish. Characterization of five forms of C3 in the diploid fish *Sparus aurata. J. Immunol.* 158:2813–2821.

Sunyer, J.O., Zarkadis, I.K., Sahu, A. and Lambris, J.D. (1996) Multiple forms of complement C3 in trout that differ in binding to complement activators. *Proc. Natl. Acad. Sci. U.S.A.* 93:8546–8551.

Suzuki, K. (1986) Morphological and phagocytic characteristics of peritoneal exudate cells in tulapia, *Oreochromis niloticus* (Trewavas) and carp, *Cyrpinus carpio. J. Fish Biol.* 29:349–364.

Sveinbjornsson, B. and Seljelid, R. (1994) Aminated β-1,3,D-polyglucose activates salmon pronephros macrophages *in vitro. Vet. Immunol. Immunopathol.* 41:113–123.

Sveinbjornsson, B., Smedsrod, B., Berg, T. and Seljelid, R. (1995) Intestinal uptake and organ distribution of immunomodulatory aminated β-1,3-D-polyglucose in Atlantic salmon (*Salmo salar* L.). *Fish Shellfish Immunol.* 5:39–50.

Tamura, E. and Honma, Y. (1970) Histological changes in the organs and tissues of gobiid fishes throughout their life-span. III. Hemopoietic organs in the ice-goby *Leucopsarion petersi* Hilgendorf. *Bull. Jap. Soc. Sci. Fish.* 36:661–669.

Tatner, M.F. (1986) The ontogeny of humoral immunity in rainbow trout, *Salmo gairdneri. Vet. Immunol. Immunopathol.* 12:93–105.

Tatner, M.F. and Manning, M.J. (1982) The morphology of the trout *Salmo gairdneri* Richardson thymus: some practical and theoretical considerations. *J. Fish. Biol.* 21:27–32.

Tatner, M.F. and Manning, M.J. (1983) The ontogeny of cellular immunity in the rainbow trout, *Salmo gairdneri* Richardson, in relation to the stage of development of the lymphoid organs. *Dev. Comp. Immunol.* 7:69–75.

Tischendorf, F. (1969) Gefässe der Milz. *In: Handbuch der Mikroskopischen Anatomie des Menschen. Blutgefäss- und Lymphgefässapparat innersekretorischer Drüsen. 6. Die Milz.* Springer-Verlag, Berlin, pp 472–654.

Trump, G.N. and Hildemann, W.H. (1970) Antibody responses of goldfish to bovine serum albumin primary and secondary responses. *Immunology,* 19:621–627.

Uemura, T., Yano, T., Shiraishi, H. and Nakao, M. (1996) Purification and characterization of the eight and ninth components of carp complement. *Mol. Immunol.* 33:925–932.

Vallejo, A.N., Miller, N.W., Jorgensen, T. and Clem, L.W. (1990) Phylogeny of immune recognition: antigen processing/presentation in channel catfish immune responses to hemocyanins. *Cell. Immunol.* 130:364–377.

Vallejo, A.N., Ellsaesser, C.F., Miller, N.W. and Clem, L.W. (1991a) Spontaneous development of functionally active long-term monocyte-like cell lines from channel catfish. *In Vitro Cell. Dev. Biol.* 27A:279–285.

Vallejo, A.N., Miller, N.W. and Clem, L.W. (1991b) Phylogeny of immune recognition: processing and presentation of structurally defined proteins in channel catfish immune responses. *Dev. Immunol.* 1:137–148.

Vallejo, A.N., Miller, N.W. and Clem, L.W. (1992a) Cellular pathway(s) of antigen processing in fish APC: effect of varying *in vitro* temperatures on antigen catabolism. *Dev. Comp. Immunol.* 16:367–381.

Vallejo, A.N., Miller, N.W., Harvey, N.E., Cuchens, M.A., Warr, G.W. and Clem, L.W. (1992b) Cellular pathway(s) of antigen processing and presentation in fish APC: endosomal involvement and cell-free antigen presentation. *Dev. Immunol.* 3:51–65.

Vallejo, A.N., Miller, N.W., Warr, G.W., Gentry, G.A. and Clem, L.W. (1993) Phylogeny of immune recognition: fine specificity of fish immune repertoires to cytochrome C. *Dev. Comp. Immunol.* 17:229–240.

van Ginkel, F.W., Miller, N.W., Cuchens, M.A. and Clem, L.W. (1994) Activation of channel catfish B cells by membrane immunoglobulin cross-linking. *Dev. Comp. Immunol.* 18:97–107.

van Muiswinkel, W.B., Lamers, C.H.J. and Rombout, J.H.W.M. (1991) Structural and functional aspects of the spleen in bony fish. *Res. Immunol.* 142:362–366.

Ventura-Holman, T., Jones, J.C., Ghaffari, S.H. and Lobb, C.J. (1994) Structure and genomic organization of VH gene segments in the channel catfish: members of different VH gene families are interspersed and closely linked. *Mol. Immunol.* 31:823–832.

Verburg-van Kemenade, B.M.L., Weyts, F.A.A., Debets, R. and Flik, G. (1995) Carp macrophages and neutrophilis granulocytes secrete an interleukin-1-like factor. *Dev. Comp. Immunol.* 19:59–70.

Ventura-Holman, T., Ghaffari, S.H. and Lobb, C.J. (1996) Characterization of a seventh family of immunoglobulin heavy chain Vh gene segments in the channel catfish, *Ictalurus punctatus. Eur. J. Immunogen.* 23:7–24.

Vilain, C., Wetzel, M.C., Du Pasquier, L. and Charlemagne, J. (1984) Structural and functional analysis of spontaneous anti-nitrophenyl antibodies of three cyprinid fish species: carp (*Cyprinus carpio*), goldfish (*Carrassius auratus*) and tench (*Tinca tinca*). *Dev. Comp. Immunol.* 8:611–622.

Walker, R.B. and McConnell, T.J. (1994) Polymorphism of the MHC Mosa class II beta chain encoding gene in striped bass *(Morone saxatilis). Dev. Comp. Immunol.* 18:325–342.

Walwig, F. (1958) Blood and parenchymal cells in the spleen of the ice fish *Chaenocephalus aceratus* (Lonnberg). *Nytt. Magasin Zool.* 6:111–120.

Wang, R. and Belosevic, M. (1994) Estradiol increases susceptibility of goldfish to *Trypanosoma danilewskyi. Dev. Comp. Immunol.* 18:377–387.

Wang, R. and Belosevic, M. (1995) The *in vitro* effects of estradiol and cortisol on the function of a long-term goldfish macrophage cell line. *Dev. Comp. Immunol.* 19:327–336.

Warr, G.W. (1995) The immunoglobulin genes of fish. *Dev. Comp. Immunol.* 19:1–12.

Warr, G.W. and Marchalonis, J.J. (1977) Lymphocyte surface immunoglobulin of the goldfish differens from its serum counterparts. *Dev. Comp. Immunol.* 1:15–22.

Warr, G.W., Miller, N.W., Clem, L.W. and Wilson, M.R. (1992) Alternate splicing pathways of the immunoglobulin heavy chain transcript of a teleost fish, *Ictalurus punctatus. Immunogenetics* 35:253–256.

Waterstrat, P.R., Ainsworth, A.J. and Capley, G. (1991) *In vitro* responses of channel catfish, *Ictalurus punctatus*, neutrophils to *Edwardsiella ictaluri. Dev. Comp. Immunol.* 15:53–63.

Weisel, G.F. (1973) Anatomy and histology of the digestive system of the paddlefish *Polyodon spathula. J. Morphol.* 140:243–256.

Weisel, G.F. (1979) Histology of the feeding and digestive organs of the shovelnose sturgeon *Scaperhynchus platorhynchus. Copeia* 518–525.

Wilson, M.R., Ross, D.A., Miller, N.W., Clem, L.W., Middleton, D.L. and Warr, G.W. (1995a) Alternate pre-mRNA processing pathways in the production of membrane IgM heavy chains in holostean fish. *Dev. Comp. Immunol.* 19:165–177.

Wilson, M.R., van Ravenstein, E., Miller, N.W., Clem, L.W., Middleton, D.L. and Warr, G.W. (1995b) cDNA sequences and organization of IgM heavy chain genes in two holostean fish. *Dev. Comp. Immunol.* 19:153–164.

Wilson, M., Bengten, E., Miller, N.W., Clem, L.W., Du Pasquier, L. and Warr, G.W. (1997) A novel chimeric Ig heavy chain from a teleost fish shares similarities to IgD. *Proc. Natl. Acad. Sci. U.S.A.* 94:4593–4597.

Winkelhake, J.L. and Chang, R.J. (1982) Acute phase (C-reactive) protein-like macromolecules from rainbow trout *(Salmo gairdneri). Dev. Comp. Immunol.* 6:481–489.

Yano, T. and Nakao, M. (1994) Isolation of a carp complement protein homologous to mammalian factor D. *Mol. Immunol.* 31:337–342.

Yocum, D., Cuchens, M.A. and Clem, L.W. (1975) The hapten/carrier effect in teleost fish. *J. Immunol.* 114:925–927.

Yoshida, S.H., Stuge, T.B., Miller, N.W. and Clem, L.W. (1995) Phylogeny of lymphocyte heterogeneity: cytotoxic activity of channel catfish peripheral blood leukocytes directed against allogeneic targets. *Dev. Comp. Immunol.* 19:71–77.

Yu, M.L., Sarot, D.A., Filazzola, R.J. and Perlmutter, A. (1970) Effects of splenoctomy on the immune response of the blue gourami, *Trichogaster trichopterus,* to infectious pancreatic necrosis (IPN) virus. *Life Sci.* 9:749–755.

Zapata, A. (1979) Ultrastructural study of the teleost fish kidney. *Dev. Comp. Immunol.* 3:55–65.

Zapata, A. (1981a) Lymphoid organs of teleost fish. I. Ultrastructure of the thymus of *Rutilus rutilus. Dev. Comp. Immunol.* 5:427–436.

Zapata, A. (1981b) Lymphoid organs of teleost fish. II. Ultrastructure of the renal lymphoid tissue of *Rutilus rutilus* and *Gobio gobio. Dev. Comp. Immunol.* 5:685–690.

Zapata, A. (1983) Phylogeny of the fish immune system. *Bull. Inst. Pasteur,* 81:165–186.

Zapata, A.G. and Cooper, E.L. (1990) *The Immune System: Comparative Histophysiology.* John Wiley and Sons, Chichester.

Zelikoff, J.T., Enane, N.A., Bowser, D., Squibb, K.S. and Frenkel, K. (1991) Development of fish peritoneal macrophages as a model for higher vertebrates in immunotoxicological studies. I. Characterization of trout macrophage morphological, functional, and biochemical properties. *Fund. Appl. Toxicol.* 16:576–589.

General conclusions

In this monograph, we have tried to show that immunity is an integral part of a general homeostatic system which helps to maintain the integrity of all multicellular organisms. Every animal is able to respond to the pathogen invasion with a cascade of orchestrated defense reactions involving neuroendocrine cooperations. To survive, an organism must be able to perform two fundamental processes: to maintain its self-integrity and to defend itself. The maintenance of integrity relies in the precise genetic plan, according to which the construction of basic body pattern is built by means of specific morphofunctional processes. The defense of the integrity has to be directed against internal and environmental factors threatening the life of an individual. The major endangering factors attacking organismic integrity and requiring homeostatic regulation are the mistakes emerging during morphogenetic growth and aging including neoplasy, parasitism and fusion with genetically unrelated organisms leading to chimaerism and resulting in loss of individuality. The transplantation reaction may be a phyletic consequence of the adaptations to the risk of that parasitism (Buss and Green, 1985). In all animal species living today the immune mechanisms evolved into an effective device which enabled their successful phylogenetic survival, and above all, their adaptive radiation in the biosphere.

At present it is clear to all scientists interested in biomedicine that since the emergence of immunology as an auxilliary branch of microbiology at the end of the 19th century, the tremendous progress of this discipline has deeply influenced directly or, in any case, at least indirectly, all biomedical sciences. When comparative and evolutionary immunology began to flower during the second half of the 20th century, it was considered to be a field, from the standpoint of other biomedical disciplines, that was exotic. It was courageous and still is, in this era of molecular sciences and reductionistic thinking, to start basic research on immune reactions of various non-traditional animal species, often phylogenetically extremely distant to man and mammals, and often without relevant tools like monoclonal antibodies or sophisticated techniques used in molecular biology and genetics. Despite the prevailing tendency of the majority of contemporary scientists to rapidly reach the most applicable results which are the most acceptable for grant agencies, basic research in comparative and evolutionary immunology has substantially contributed in many ways to the exciting progress that immunology has made in recent decades. It must be admitted that molecular biology transformed immunology, so that the mechanics of function of the immune system are more understandable, nevertheless mankind is still plagued by many immunological disorders the elucidation of which may help the investigations of other immune strategies used by non-mammalian animals.

Myriads of species on Earth have already succumbed to extinction. Some paleontologists figure that from at least 5 to 50 billion of various species, more than 99% have vanished. We rarely study or think about that dissappearance of species, which according to new paleontologic discoveries and confirmation of older ones have been a regularly occurring phenomena since the Phanerozoic period and the emergence of multicellular animals. In reality, the anagenetic evolution is not only a simple pathway from less complex to highly hierarchized organisms, but it went through confusing stages when more advanced groups could be overcome by primitive taxa for a couple of periods, which again ultimately went extinct.

Our vision of the evolutionary processes which have been interpreted in the past as every phylum is „better" than the previous one, is a great mistake. Similarly, we must admit that adaptive immunity has not necessarily been an ideal solution for survival in nature. Many of the present living creatures utilize simple immune strategies and still survive very successfully together with mammals. Yet the causes behind those great extinctions remain unclear. Numerous theories try to explain what events could have led to extinctions, but only rarely is the problem interpreted by considering immunology. It is certain that whatever caused vanishing of great animal assemblages, such as rapid oscillations of earth's climate and/or poor endowment of genome including accumulations of fetal mutations, and furthermore, owing to that no extinction has occurred rapidly but rather within longer time spans, the final act of dying out might be caused by epidemics as a consequence of insufficiently developed immune potential which was unable to vanguish new infectious diseases (Raup, 1981). For the time being, the available data obtained by extensive study of immune reactions of non-mammalian animals leave the question of the role of immunity in the extinction or survival of species still open.

The history of mankind has been modified by epidemic diseases. How important the immune system is for our survival can be continually demonstrated in cases when we are unable to erradicate infectious diseases and parasites which still kill millions of humans, such as influenza and malaria, or when new diseases like AIDS emerge. In this connection, it seems appropriate to note that the contribution of comparative immunology is obvious from the point of view of longer-range perspectives of clarification of such problems.

It is generally believed that the evolution of the central nervous system could be one of the main propellers of progress in phylogeny. The brain is considered to be the most complex structure in the universe. But we should not forget that our immune system, composed at least of 10^{12} of cells, is in its complexity of structure and function comparable to CNS. If immunity has played a greater role in evolutionary alterations of species, and this cannot be excluded according to the new discoveries, then comparative and evolutionary immunology studies must focus ont the immune strategies which are utilized at present by various living animal species and which have emerged during evolution from different ancestral backgrounds.

Contemporary knowledge summarized by evolutionary and comparative immunology represents only the tip of an iceberg. We are only now arranging our knowledge in this field. We now address enormous tasks and questions: how is it possible that more simple and often seemingly non-effective immune endowment of some animal taxa allowed their survival for certain periods? How is it possible to battle pathogens and parasites threatening their organismic integrity without relevant recognition receptors or specific defense and regulatory macromolecules? In what mechanism is hidden the secret of their evolutionary successful survival? We feel that the answers to these questions will bring exciting discoveries with more than theoretical importance.

To overlook or underestimate any new information which this fruitful branch offers would limit our scientific creativity and deprive us of the opportunity to discover something new. It would not be prudent to neglect any possibility which could contribute to the benefit of mankind.

References

Buss, L. V. and Green, D. R. (1985) Histocompatibility in vertebrates: the relict hypothesis. *Dev. Comp. Immunol.* 9:191–201.

Raup, D. M. (1981) Extinction: bad genes or luck? *Acta Geol. Hisp.* 16:25–33.

Index

Toxicology • Ecology • Zoology

EXS 86

T. Braunbeck, Institute of Zoology, University of Heidelberg, Germany
B. Streit, Institute of Zoology, University of Frankfurt, Germany
D. Hinton, School of Veterinary Medicine, University of California at Davis, USA (Eds)

Fish Ecotoxicology

1998. Approx. 300 pages. Hardcover.
ISBN 3-7643-5819-X
Due in September 1998

In the last twenty years, ecotoxicology has successfully established its place as an interdisciplinary science concerned with the effects of chemicals on populations and ecosystems, thus bridging the gap between biological and environmental sciences, ecology, chemistry and traditional toxicology. In modern ecotoxicology, fish have become the major vertebrate model, and a tremendous body of information has been accumulated. This volume attempts to summarize our present knowledge in several fields of primary ecotoxicological interest ranging from the use of (ultra)structural modifications of selected cell systems as sources of biomarkers for environmental impact over novel approaches to monitoring the impact of xenobiotics with fish in vitro systems such as primary and permanent fish cell cultures, the importance of early life-stage tests with fish, the bioaccumulation of xenobiotics in fish, the origin of liver neoplastic lesions in small fish species, immunocytochemical approaches to monitoring effects in cytochrome P450-related biotransformation, the impact of heavy metals in soft water systems, the environmental toxicology of organotin compounds, oxidative stress in fish by environmental pollutants to effects by estrogenic substances in aquatic systems. This collection of up-to-date reviews thus covers a broad range of topics important in current ecotoxicological research and should be of interest not only to specialists in ecotoxicology, but also to scientists and students in related fields of environmental sciences.

For orders originating from all over the world except USA and Canada
Birkhäuser Verlag AG
P.O. Box 133
CH-4010 Basel / Switzerland
Fax: +41/61/205 07 92
e-mail: orders@birkhauser.ch

For orders originating in the USA and Canada
Birkhäuser
333 Meadowland Parkway
USA-Secaurus, NJ 07094-2491
Fax: +1 201 348 4033
e-mail: orders@birkhauser.com

Birkhäuser

Evolutionary Biology • Population Biology • Ecology

EXS 83

R. Bijlsma, University of Groningen, The Netherlands
V. Loeschcke, University of Aarhus, Denmark (Eds)

Environmental Stress, Adaptation and Evolution

1997. 344 pages. Hardcover.
ISBN 3-7643-5695-2

Most organisms and populations have to cope with hostile environments, threatening their existence. Their ability to respond phenotypically and genetically to these challenges and to evolve adaptive mechanisms is,
therefore, crucial.

The contributions to this book aim at understanding, from an evolutionary perspective, the impact of stress on biological systems. Scientists, applying different approaches spanning from the molecular and the protein level to individuals, populations and ecosystems, explore how organisms adapt to extreme environments, how stress changes genetic structure and affects life histories, how organisms cope with thermal stress through acclimation, and how environmental and genetic stress induces fluctuating asymmetry, shapes selection pressure and causes extinction of populations. Finally, it discusses the role of stress in evolutionary change, from stress-induced mutations and selection to speciation and evolution at the geological time scale.

The book contains reviews and novel scientific results on the subject. It will be of interest to both researchers and graduate students and may serve as a text for graduate courses.

For orders originating from all over
the world except USA and Canada
Birkhäuser Verlag AG
P.O. Box 133
CH-4010 Basel / Switzerland
Fax: +41/61/205 07 92
e-mail: orders@birkhauser.ch

For orders originating in the
USA and Canada
Birkhäuser
333 Meadowland Parkway
USA-Secaurus, NJ 07094-2491
Fax: +1 201 348 4033
e-mail: orders@birkhauser.com

Evolutionary Biology • Ecology • Limnology

EXS 82

B. Streit, University of Frankfurt, Germany
T. Städler, University of Frankfurt, Germany
C.M. Lively, Indiana Univ., Bloomington, IN, USA (Eds)

Evolutionary Ecology of Freshwater Animals
Concepts and Case Studies

1997. 384 pages, Hardcover.
ISBN 3-7643-5694-4

Evolutionary ecology includes aspects of community structure, trophic interactions, life-history tactics, and reproductive modes, analyzed from an evolutionary perspective. Freshwater environments often impose spatial structure on populations, e.g. within large lakes or among habitat patches, facilitating genetic and phenotypic divergence. Traditionally, freshwater systems have featured prominently in ecological research and population biology.

This book brings together information on diverse freshwater taxa, with a mix of critical review, synthesis, and case studies. Using examples from bryozoans, rotifers, cladocerans, molluscs, teleosts and others, the authors cover current conceptual issues of evolutionary ecology in considerable depth.
The book can serve as a source of critically evaluated ideas, detailed case studies, and open problems in the field of evolutionary ecology. It is recommended for students and researchers in ecology, limnology, population biology, and evolutionary biology.

For orders originating from all over   For orders originating in the
the world except USA and Canada   USA and Canada
Birkhäuser Verlag AG   **Birkhäuser**
P.O. Box 133   333 Meadowland Parkway
CH-4010 Basel / Switzerland   USA-Secaurus, NJ 07094-2491
Fax: +41/61/205 07 92   Fax: +1 201 348 4033
e-mail: orders@birkhauser.ch   e-mail: orders@birkhauser.com

Evolutionary Biology

R. deSalle, American Museum of Natural History, New York, USA
B. Schierwater, University of Frankfurt, Germany (Eds)

Molecular Approaches to Ecology and Evolution

1998. Approx. 275 pages. Hardcover.
ISBN 3-7643-5725-8
Due in September 1998

The last ten years have seen an explosion of activity in the application of molecular biological techniques to evolutionary and ecological studies. This volume attempts to summarize advances in the field and place into context the wide variety of methods available to ecologists and evolutionary biologists using molecular techniques. Both the molecular techniques and the variety of methods available for the analysis of such data are presented in the text. The book has three major sections - populations, species and higher taxa. Each of these sections contains chapters by leading scientists working at these levels, where clear and concise discussion of technology and implication of results are presented.
The volume is intended for advanced students of ecology and evolution and would be a suitable textbook for advanced undergraduate and graduate student seminar courses.

For orders originating from all over the world except USA and Canada
Birkhäuser Verlag AG
P.O. Box 133
CH-4010 Basel / Switzerland
Fax: +41/61/205 07 92
e-mail: orders@birkhauser.ch

For orders originating in the USA and Canada
Birkhäuser
333 Meadowland Parkway
USA-Secaurus, NJ 07094-2491
Fax: +1 201 348 4033
e-mail: orders@birkhauser.com